Sustainable Development Goals Series

The **Sustainable Development Goals Series** is Springer Nature's inaugural cross-imprint book series that addresses and supports the United Nations' seventeen Sustainable Development Goals. The series fosters comprehensive research focused on these global targets and endeavours to address some of society's greatest grand challenges. The SDGs are inherently multidisciplinary, and they bring people working across different fields together and working towards a common goal. In this spirit, the Sustainable Development Goals series is the first at Springer Nature to publish books under both the Springer and Palgrave Macmillan imprints, bringing the strengths of our imprints together.

The Sustainable Development Goals Series is organized into eighteen subseries: one subseries based around each of the seventeen respective Sustainable Development Goals, and an eighteenth subseries, "Connecting the Goals," which serves as a home for volumes addressing multiple goals or studying the SDGs as a whole. Each subseries is guided by an expert Subseries Advisor with years or decades of experience studying and addressing core components of their respective Goal.

The SDG Series has a remit as broad as the SDGs themselves, and contributions are welcome from scientists, academics, policymakers, and researchers working in fields related to any of the seventeen goals. If you are interested in contributing a monograph or curated volume to the series, please contact the Publishers: Zachary Romano [Springer; zachary.romano@springer.com] and Rachael Ballard [Palgrave Macmillan; rachael.ballard@palgrave.com].

Flora Lu • Emily Murai
Editors

Critical Campus Sustainabilities

Bridging Social Justice and the Environment in Higher Education

Springer

Editors
Flora Lu
University of California
Santa Cruz, CA, USA

Emily Murai
University of California
Santa Cruz, CA, USA

ISSN 2523-3084 ISSN 2523-3092 (electronic)
Sustainable Development Goals Series
ISBN 978-3-031-30928-1 ISBN 978-3-031-30929-8 (eBook)
https://doi.org/10.1007/978-3-031-30929-8

This Springer imprint is published by the registered company Springer Nature Switzerland AG
The registered company address is: Gewerbestrasse 11, 6330 Cham, Switzerland

Paper in this product is recyclable.

To the members of the campus community and broader community for whom environmental concerns were always inextricable from demands for respect, recognition, and repairing relations

Preface

We undertake the writing and editing of this book as two Asian American women sharing many commonalities in our academic interests, institutional affiliations, interdisciplinary training, and work styles. We are both faculty in the department of Environmental Studies at the University of California, Santa Cruz (UCSC), and also serve as leaders in the campus' college system (namely College Nine and John R. Lewis College). Emily is an alumna of UCSC, earning her BA in Psychology and American Studies before heading off to graduate school at the New School for Social Research (MA, Cultural Sociology) and a PhD in Geography from the University of Minnesota. Flora earned her AB in Human Biology from Stanford and her PhD in Ecology from University of North Carolina at Chapel Hill. We are both passionate about working with students, creating welcoming academic spaces, and undertaking innovative pedagogical approaches that draw connections between social justice and environmental sustainability. As daughters of immigrants, the messages of keeping your head down, working extremely hard, and going above and beyond for the collective good were received loud and clear. In this preface, we wanted to share a little about our backgrounds and lived experiences, and why the topics we cover in this edited book resonate so deeply with us.

The insights of this volume cannot be extricated from the disruptive, tectonic jolts that have characterized these past six years, from racial reckonings and climate catastrophes, to economic emergencies and pandemic privation. The datasets we share in this volume begin in 2016, with the first People of Color Sustainability Collective campus-wide survey implemented between Winter and Spring quarters. The presidential election later that year, and the subsequent mainstreaming of white supremacist, xenophobic, misogynistic, and authoritarian dogmas, behaviors and rhetoric, were profoundly felt even in our (supposedly) progressive bastion of UCSC. Trump's example, and the inescapable media saturation of his daily assaults on decency, created a permission structure, and normalized the abnormal, muted what used to be alarming, and magnified Alt-Right ideology. Students picked up on these messages and in spite of—or perhaps because of—UCSC's status as a Hispanic-Serving Institution (HSI) and Asian American and Native American Pacific Islander-Serving Institution (AANAPISI), students who identified with conservative groups began to make farcical claims of being marginalized and victimized on campus. Racist and hateful fliers and graffiti started to appear on campus. Students in our classes openly challenged our authority, questioned our "facts," and dismissed lessons we were trying to teach. Students of color, undocumented students, and students from other marginalized groups were fearful and in retreat in this political climate.

Events like these have pushed the bounds of scholarship, generating new insights, expanding concepts, and creating moments to re-theorize and re-analyze taken-for-granted concepts. The need to reconsider mainstream environmental and sustainability concepts and practices has taken on renewed urgency given the way non-human nature—through extreme climate events, devastating wildfires, and global pandemics that were once considered "once-in-a-lifetime"—has repeatedly asserted itself over the last few years despite attempts by the capitalist-politico-industrial complex to downplay and minimize its effects. This window of time is thus important to understand from the perspective of students struggling to make sense and meaning of the world around them, as heirs to the social and environmental issues we pass on to them. That is why, in various chapters, we share vignettes of moments of students' agency and activism, and the ways they made connections between personal and political, environmental and epidemiological, lives and livelihoods. They are taking it upon themselves to improve the future for their own longevity.

The struggle to make sense out of these issues and understand their impacts on environmental issues is not limited to students. They seep into and shape the lives of faculty and staff within the academic community as well. For us—as Chinese/American and Chinese/Japanese/American women in a mainstream environmentalist-oriented department dominated by white men (as most academic departments are; our department is not unique in that white men are over-represented at the most senior levels)—these last half a dozen years have brought into stark relief the challenges ahead. We share our experiences not as an indictment of others in our midst, but as a way of bringing to light what has been hidden, elided, and marginalized for too long.

After the election of 2016, a leader in the department (a white male natural scientist) stymied efforts to draft and disseminate a statement in support of students—queer, undocumented, BIPOC (Black, Indigenous, and People of Color), and of other marginalized identities—who were afraid and upset. Other departments and units across the university were sharing such public declarations of solidarity and care. Doing so, he claimed, would not be respectful to conservatives and Christians on campus—an odd claim given the overwhelming representation and power these groups have in American society more generally. Again in 2019, this same faculty member complained about an anti-Trump sign that some graduate students in another department put on the door of their office, located in the same building as the Environmental Studies Department. He said it was because it was not in accordance with UCSC's Principles of Community (University of California, Santa Cruz, 2017) and not welcoming of those of "conservative political views," "republican family background," and "deeply held religious beliefs." The sign, which said "Destroy Trump" and "BOOOOOO" surrounded by smiley faces and one with a tongue sticking out, was to this professor something that would "suggest that the department [with which these graduate students are affiliated] is hostile to the point of personal violence toward anyone who does not share the political views of the person who wrote the sign." The chair of the department with which these graduate students were affiliated responded to the Environmental Studies faculty member's complaint, stating that the students were drawing attention to policies of oppression enacted by the Trump administration, writing that protests "are meant to galvanize the power of the dispossessed and put the powerful on notice that the status quo cannot persist."

The actions of the senior Environmental Studies faculty member exemplifies "a larger neoliberal trend that paints diversity as a beneficial resource to which all may lay equal claim, unmooring it from unequal distributions of power. Forget about rectifying the legacy of inequality and racial violence, this rationale suggests. Everyone is included, even if everyone is not equally marginalized" (Asultany, 2022). For those who enjoy racial, gender, economic, and other forms of privilege to insist that those with less privilege tolerate systems of oppression, and to marshal discourses of diversity and inclusion in doing so, reflect at the very least a failure to understand concepts such power, positionality, and reflexivity. This can be a shocking revelation to those of us who see these concepts as foundational to our lives and work and expect campus leaders to be attuned to these ideas in their roles as leaders—how can they weigh-in on complex, intersecting and controversial issues without this understanding? How can they support and protect colleagues and community members with different identities and backgrounds without sensitivity to the obstacles they face? Campus leaders who gloss over or trivialize these issues, or worse, claim that those with socially privileged identities are the ones who are, in fact, victimized, create a toxic and hostile work environment by silencing and gaslighting those of us who experience multiple levels of marginalization, while we exhaust ourselves in the push to be legitimated and seen.

About a year after the Trump protest sign issue, the UCSC Environmental Studies Department was the site of an incident of hate speech on social media that garnered national attention, speech that was targeted against people of Asian background and especially those from China. The post, reported to have been written by a then-graduate student in Environmental Studies, disparaged Chinese culture, reiterated Trump's COVID-19 anti-Asian slurs, and de-humanized Chinese people. Former President Trump often made reference to the "China Virus" and the "Kung Flu," indifferent to the ways this rhetoric legitimated and promoted acts of racially motivated harm and violence against innocent people. In the wake of the COVID-19 pandemic, a spate, still unabated, of violence has targeted people of Asian descent in this country, evoking historical tropes of Asians as predatory, disease-ridden, perpetual foreigners. Recent reporting from the *LA Times* and *SF Chronicle* reported that anti-AAPI hate crimes in California jumped 177% in 2021, and that San Francisco saw more than a six-fold increase during the same time. The organization Stop AAPI Hate documented almost 11,000 instances of anti-Asian hate incidents between March 2020 and December 2021. AAPI women are twice as likely to report hate incidents than men, such as verbal harassment, shunning, being spat on, workplace discrimination, vandalism, and physical assault. The National Asian Pacific American Women's Forum found that 78% of AAPI women reported being affected by anti-Asian racism between 2019–2021 and 55% had personally encountered specific instances of racism.

Against this backdrop, this incident of hate speech caused despair, fear, worry, confusion, hurt, and anger among students, staff, and faculty in the Environmental Studies Department and beyond. Undergraduates shared that they were "disappointed and disheartened to be an Asian person in the ENVS major," and that their parents in China were concerned for their safety. Another person stated,

> Right now, we are known across campus and beyond for being a space of anti-AAPI sentiment…People are deeply upset and triggered. Yet literally nothing is being done. I asked the current chair to make a statement after the hate crime in GA [in which eight people at massage businesses in Atlanta were killed in mid-March 2021], and he did not…The faculty deeply need to have serious dialogue about what the hell is going on. It cannot be predominantly women of color doing the heavy lifting to address racism in ENVS…we need emotional support for the people who are hurting. But instead, they continue to be hurt. (Environmental Studies DEIA Survey Report 2022: 38)

In response to the post, the ENVS Diversity Committee and the ENVS graduate students both released statements calling for a fundamental review and revision of curricula in the department and sustained and meaningful anti-racism efforts. To cite the latter ("ENVS Graduate Student Commitment to a Diverse, Anti-Racist Department," April 10, 2020):

> As a graduate student community, we are committed to dismantling narratives like these and the violent practices associated with them. Many of us are working to build new traditions of scholarship that confront the political-economic and white supremacist structures that underpin environmental challenges and injustices. As researchers, we strive to practice reflexivity, acknowledge our positionality relative to the people involved in and impacted by our work, and call out modern scientific practices that exploit, exclude, and marginalize non-white, non-Western peoples and knowledges.
>
> We see this alleged incident not as an isolated occurrence, but as a flashpoint for broader concerns about racism and colonialism in our department and the academy. We pledge to support efforts to find a path forward that prioritizes the needs of students of color and the university's mission to improve campus climate. Clearly, this must include:
>
> - Critical evaluation of undergraduate and graduate curriculum in Environmental Studies.
> - Additional training for Environmental Studies faculty and graduate students, which could draw on available resources like the ODEI's (Office of Diversity Equity and Inclusion) Intersections between Diversity and Environment short course.
> - Dialogue between the administration, graduates, and undergraduate students to identify additional steps. For example, the interactive poster in the 4th floor ISB hallway titled "Envisioning an anti-racist ENVS department" suggests hiring Black and Latinx faculty and creating people-of-color-centered syllabi.
> - Honest discussions about the accessibility of higher education, including graduate students' demands for a living wage and undergraduates' demands for tuition reduction.

In spring of 2021, the UCSC Environmental Studies Diversity Committee—of which Flora was founding chair and Emily was a member—developed and undertook a survey of our department's graduate students, faculty, and staff to gauge their opinions, feelings, and experiences with diversity, equity, inclusion, and anti-racism (DEIA) issues, partially out of response to the incident. The survey was prepared in the spirit of vulnerability, transparency, and authenticity in the hopes that it would "establish a stronger baseline of knowledge from which to continue making progress in our DEIA efforts, help to center these goals within our department's core competencies, and foreground an ethos of respect, reflexivity, and care that will sustain our community members, individually and collectively."

The following year, the Diversity Committee released a report detailing the findings of the survey. Of the 43 respondents (out of an estimated total of ~100 potential respondents), some patterns in the findings include a "deeply insufficient understanding in the department of institutional racism and intersectionality," which led to 90.2% of respondents selecting race as a key priority area for ENVS to address; a privileging of natural sciences over social sciences, which some respondents felt had racial implications; and an overall departmental discomfort with advising students from diverse backgrounds and discussing issues of diversity. On identifying recommendations and ways forward, the report made suggestions such as the hiring and retention of more BIPOC staff and faculty; making sure every graduate student has a rigorous grounding in DEIA issues; improving pedagogical training for faculty and graduate students; providing neutral party mediation and support services outside of the department in cases of grievance; better mental health support; develop a strategic plan for DEIA in the department; and recognize and reward DEIA efforts. Yet, despite having a comprehensive document of feedback and concrete suggestions, the department has not made any notable progress in working toward these outcomes, raising questions as to whether it sees these changes as necessary or important, and when change, if ever, will occur.

Clearly, the issues, topics, and debates discussed in this edited volume around critical sustainability and environmentalism, social justice, equity, and needed changes in institutions of higher education are not merely intellectual interests of ours. They are deeply personal and inextricable from our lived experiences as scholars, mentors, teachers, and members of a campus community. And just as the graduate students in Environmental Studies held some of the sharpest and most salient insights and suggestions, so do our undergraduates when it comes to what our campus needs to be doing differently. In this book, we have tried to center the voices of our students, current and former, who call on us to do more and to do better, from learning with and from community partners to increasing spaces and expectations for faculty of all disciplines to be knowledgeable about and work toward critical sustainabilities.

Santa Cruz, CA, USA

Flora Lu
Emily Murai

Acknowledgments

We recognize that academic and co-curricular work relies on the goodwill, patience, generosity, and trust of a web of collaborators, supporters, and partners. The authors would like to thank community leaders and collaborators who educate us, inspire us, and remind us of what is important: the families of Calabasas Community Garden, Principal Todd Westfall of Calabasas Elementary School; Chairman Valentin Lopez of the Amah Mutsun Tribal Band; MariaElena de La Garza of the Community Action Board of Watsonville; Carmen Herrera-Mansir, Cesario Ruiz and Amy Mascareñas of the El Pájaro Community Development Corporation; Roberta Ruiz-Comacho of Gonzales; Lyla June; Daniel Nane Alejandrez of Santa Cruz Barrios Unidos; and Mark and Barbara Gordon of the Osher Lifelong Learning Institute. We express our deep appreciation to the Ethnic Resource Centers: Nancy Kim, Caz Salamanca, Shonté Thomas, Travis Becker, Cameron de Leon, Autumn Johnson, Angel Riotutar, Jemzi Ortiz, and Judith Estrada. Sincere thanks to the staff of College Nine and John R. Lewis College: Sarah Woodside Bury, Wendy Baxter, Deana Slater, Robin Kirksey, Denise Booth, Madison Armstrong, Audrey Kim, Erin Ramsden, Mark Gardner, Gabie Darlak, Angela Birts, Crystal Knight, Daphne Mark, Lenora Willis, Tim Barbour, Christian Galliguez, Jessica Lawrence, Abbey Asher, and Jane Hartman. Gratitude to the staff of the Sustainability Office, especially Alessandra Alvares, Kristen Lee, Shauna Casey, Jessica Zubia Calsada, Ellen Vaughn, and Ileana Brunetti. Thank you to our many other colleagues and friends: John Brown Childs, Lisa O'Connor, Stacy Philpott, Ravi Rajan, Sikina Jinnah, Jeff Bury, Steve McKay, Chris Benner, Madeleine Fairbairn, Maywa Montenegro de Wit, Carol Shennan, Christine Hong, Xavier Livermon, Amy Ginther, Christy Byrd, Sylvanna Falcon, Derede Arthur, Helen Shapiro, Cam Leaper, Samuel Severance, Aysha Peterson, Justin Luong, Clara Qin, Alejandro Artiga-Purcell, Rick Flores, Tracy Liu, Sheeva Sabati, Vivian Underhill, Betsy Lopez-Wagner, Sheridan Noelani Enomoto, John Mataka, Rosenda Mataka, and Dalila Adofo.

The People of Color Sustainability Collective has been funded by the UCSC Student Fees Advisory Committee (SFAC), Campus Provost/Executive Vice Chancellor Marlene Tromp, Campus Provost/Executive Vice Chancellor Alison Galloway, an anonymous donor, and the Institute for Social Transformation. The Apprenticeship in Community Engaged Research [(H) ACER] Program recognizes support from SFAC, Measure 43, the Spencer Foundation, National Science Foundation, Belmont Forum and NORFACE,

Kathleen Rose, and the US Department of Agriculture Hispanic Serving Institutions Education Program.

We extend appreciation to the Springer Nature editorial staff, Herbert Moses, Aaron Schiller, A. Meenahkumary, Margret Joy, and Henry Rodgers for their patience and support in pushing this project along.

Finally, we want to sincerely thank all the students with whom we have had the pleasure of working with, mentoring, organizing with, teaching, and learning from: our PoCSC interns, AIRC interns, (H)ACER students, CUIP interns, College Nine and JRLC student leaders and undergraduate affiliates, Environmental Justice Youth Leadership Academy Students, Alternative Spring Break participants, PRAXIS students, and undergraduate advisees. You are the inspiration for this book and the work behind it.

References

Alsultany, E. (2022). Diversity defanged: In higher ed, the concept has lost all meaning. https://www.chronicle.com/article/diversity-defanged

University of California, Santa Cruz. (2017). *Principles of community*. Retrieved January 31, 2023, from https://www.ucsc.edu/about/principles-community.html

Contents

Introduction

Flora Lu, Emily Murai, and Christopher Lang

Abstract Students across campuses are calling for sustainability education and action that is equity-centered, liberatory, culturally grounded, anti-racist, decolonial, and collective. Sustainability has traditionally been approached as a solely environmental project focused on individual consumer behaviors and techno-fixes. However, contemporary debates about sustainability are significantly more expansive and are the field through which diverse priorities and visions for the future are asserted and contested. In this chapter, we articulate our framework of critical sustainabilities in higher education, which is committed to: (1) disrupting normative narratives and epistemic hierarchies; (2) elevating relationality and connection, between people and place, with each other and non-humans; (3) confronting settler colonialism and white supremacy in higher education; and (4) centering collective care.

Keywords Traditional sustainability, critical sustainabilities, three-legged stool of sustainability, environmental justice, higher education, relationality, collective care

On a rainy afternoon in mid-February 2019, about two dozen people, almost all students, met in a room at Rachel Carson College at the University of California, Santa Cruz (UCSC). The room was organized into two groups of chairs organized in semi-circles in front of two screens, mirroring each other. On one side, the topic area was "Communal/Personal" Sustainability and the other was "Education/Academics." This was one of four "Blueprint Breakouts" that quarter, organized by the Student Environmental Center (SEC), advertised as a space "where students, faculty, alumni, and staff come together to create visions and actions for a more just and sustainable campus!" Taken together, these breakout sessions provided material for the Blueprint for a Sustainable Campus (BSC), a visioning document that has existed at UCSC for two decades, a document that is the result of collective discussions in which student voices are centered.

F. Lu · E. Murai · C. Lang
University of California, Santa Cruz, CA, USA

In the Education/Academics group, there was one staff member from the Sustainability Office (Shauna Casey) and one faculty member (Flora Lu) in attendance. The remaining ten people were undergraduates, mostly in their third or fourth year, from a variety of majors (Environmental Studies, Critical Race and Ethnic Studies, Sociology, Linguistics, Biology, Economics, Theater Arts, Psychology), the majority identifying as female. As is common among student gatherings, the discussion opened by everyone briefly introducing themselves and then reviewing community agreements (e.g., don't interrupt each other, step back if you are talking too much, step up if you are not talking, assume good intentions, listen actively, critique the idea and not the person, etc.).

Two students then presented a PowerPoint to provide necessary background, one that highlighted issues of educational equity, the centrality of environmental justice (EJ) in sustainability, and the importance of representation in terms of diverse faculty, curricula, access to education, and student voice. The PowerPoint ended with a series of questions to spark discussion, from inviting folks to share their experience with environmental sustainability education to asking them how EJ could be incorporated more into the curriculum. The conversation on the other side of the room, about Individual/ Communal Sustainability, proceeded in a similar fashion, with community agreements, a brief PowerPoint, and open discussion on the topic of how the work of sustainability on campus must be predicated upon the ability of students to meet their basic needs.

The topic areas of these Breakouts bear little resemblance to the discussions upon which earlier Blueprints for a Sustainable Campus were based. The first edition of the BSC was in 2003–2004; it was a document organized into eight topic areas: campus food systems; transportation; long-range plan (i.e., land use and campus growth); curriculum; waste prevention; green purchasing; green building and renewable energy; and campus ecosystem preservation. While undoubtedly important, these issues focus on resource management and techno-infrastructural efforts, clearly the predominant approach in sustainability policies both then and now. Discussion of race, class, and equity is largely absent; the term "social justice" appears once in the 33-page document, and the word "culture" only appears when preceded by "agri," "perma," or "consumer."

The most recent edition (2019–2021) of the BSC breaks down sustainability into nine topic areas: institutional change and UCSC; communal/personal sustainability; transportation and housing; natural resources; education/academics; food sovereignty; waste; green purchasing and building; and energy and technology. Although the topic areas remain somewhat unchanged, the content and intention of the BSC has. For the first time, environmental justice, and not just sustainability, was the central lens of the document. In the 21-page document, "justice" appears 62 times. "Today we recognize that true sustainability cannot be attained without equity, sovereignty, and justice. So in order for our sustainability plans to be aligned with that understanding, the Blueprint must, therefore, assess the campus's sustainability through a lense [sic] of environmental and social justice…[through] the 17 Principles of Environmental Justice" (BSC, 2019–2021: 5). Although the BSC has included a section on social and environmental justice since 2006, this version imbues these considerations as fundamental to BSC's visions and actions. The 2019–2021 BSC calls for fundamental changes in faculty diversity, sustainability curricula, conceptualizations of "the environment", and broader EJ literacy. It states:

> Especially in a time defined by climate catastrophe and intense political polarization, marginalized communities face the brunt of human-induced climate change but are left out of necessary conversations. Despite priding itself on being one of the top research universities in the state, academics at UC Santa Cruz continually negate the importance of social justice, environmental justice, and inclusivity within the sustainability movement. (BSC, 2019–2021: 26)

In the "Visions and Actions" sections for both the Communal/Personal Sustainability and Education/Academics sections in the BSC, the onus is on the university—faculty, administrators, and staff—to make changes that acknowledge the difficulties that UCSC students face. These difficulties include a lack of affordable housing, food security, sense of belonging and well-being, all of which undermine the ability of students to have the bandwidth to dedicate themselves to being positive change agents. In addition to promoting a broad definition of health and well-being, sustainability is about accountability, democratic and transparent decision-making processes, participation, and increased opportunities for experiential and community-engaged learning. "Doing" sustainability at UCSC means grappling with disparities of power, knowledge, access, and representation, where students, staff, and faculty have less hierarchical relationships and are intentional about equity efforts.

In the past few years, there has been a vociferous call by BIPOC (Black, Indigenous, and People of Color) students at universities to fundamentally center justice, equity, and inclusion in environmental education, programming, and institutions in higher education. For example, in March of 2017, a "Letter to the Environmental Community" was released at UC Berkeley with the hashtag #EnvironmentalismSoWhite that reads:

> This academic institution does not validate our existence, our power nor our unique knowledge in environmental dialogue. This is evident when asked to separate lived experiences of environmental degradation from academic perspectives on environmental issues, which uphold the notion that access to a clean and sustainable environment is reserved for the white and/or affluent. We [students of color at the university] have been repeatedly marginalized and invalidated in spaces where we should have felt welcomed and respected. We have had students come to us with stories that they have experienced microaggressions, cultural appropriation, tokenization, and racism from some ECO-affiliated organizations. Students of color are continuously leaving the environmental community due to feelings of alienation and frustration. We are disturbed by the lack of support and inclusion for students of color in the environmental community despite the fact that our participation is critical to the survival of your movement.

In 2020, the National Academies of Sciences commissioned a study entitled "Strengthening Sustainability Programs and Curricula at the Undergraduate and Graduate Levels." Just as the authors note the significant increase in US degree programs, research institutes and centers focused on sustainability, they also lament the lack of students from non-dominant communities in these units. Framing sustainability programs in more inclusive ways and correcting misperceptions (e.g., that BIPOC are disinterested in the environment) are necessary in striving toward greater diversity, equity and inclusivity in sustainability education. As noted in the report, "Students graduating with sustainability degrees should have an understanding of the role that processes such as colonialism, conquest, land ownership and appropriation, segregation and racism play in contemporary sustainability challenges" (NAS, 2020: 140).

The COVID-19 pandemic has laid bare the inequities at the intersections of race, ethnicity, class, immigration status, environmental injustice, and community health. More so than ever, students at institutions of higher learning are demanding curricula that center environmental equity and a diversity of ways of knowing and stewarding the planet. Scholarship is exposing histories of white supremacy and settler colonialism in environmental programs and the larger university. And shortfalls in students' basic needs, from food insecurity to high rent burden, are being framed as manifestations of environmental injustice. We have witnessed a burgeoning movement among universities seeking to infuse sustainability efforts with social justice. Often these are predominantly white institutions seeking to enrich and expand an established sustainability agenda to reflect and respond to societal shifts. Campuses are responding with anti-racist and social equity messages in a variety of ways, such as in their departmental program statements, in land acknowledgements dutifully performed before events, and within diversity training for predominantly white environmental faculty. Yet, these efforts cannot and do not reconcile the incommensurability between the campus' stated dedication to ideals of environmentalism and diversity paired with the way that students experience these ideals as actual practices.

This brings us back to the example of the Blueprint for a Sustainable Campus. The authors note that there is much more work to do:

> We will not suddenly go from a "white" organization to an organization that better reflects our community. We are nevertheless working to change the way that we think and act in order to include people from different backgrounds and experiences and *to draw from the knowledge of campus organizations such as the People of Color Sustainability Collective. We recognize PoCSC as an organization that supports students in using their knowledge and agency to change the concept of sustainability.* (BSC, 2019–2021: 7, emphasis added)

In this edited volume, we aim to make a broader call for higher education leaders to think more deeply about sustainability programming, pedagogy, research—multiple axes of sustainability praxis—by showcasing how we mobilize student and community perspectives in our research, teaching, programming to unsettle ossified and vacant conceptualizations of sustainability, both materially and discursively. The chapters that follow focus mainly, but not exclusively, on the People of Color Sustainability Collective, which from its inception centered issues of race, decoloniality, justice, and equity to address environmental challenges.

The People of Color Sustainability Collective

Founded in 2015, the People of Color Sustainability Collective (PoCSC) is a groundbreaking initiative between the Ethnic Resource Centers; Sustainability Office; and College Nine and John R. Lewis College (whose themes are "International and Global Perspectives" and "Social Justice and Community," respectively). The administrative home of PoCSC is the Ethnic Resource Centers (ERCs), specifically the American Indian Resource Center, reflecting a commitment to anti-racist and anti-colonial work, and recognizing the ERCs' pivotal role in PoCSC's establishment. The need for the initiative emerged as students of color reported experiencing sustainability-related microaggressions that undermined their sense of belonging at the institution. White, relatively wealthy students are widely perceived to dominate the environmental movement on campus, while people of color and low-income students need to be "educated" in sustainability, their environmental efforts (e.g., reusing, reducing consumption, repurposing, limiting waste, etc.) discounted, and their consumption patterns criticized. PoCSC emerged out of a confluence of factors that characterize many educational institutions: (1) a noted deficiency in campus sustainability plans, which identified technical and operational goals for the campus, but included very few targets to address issues related to justice and/or inclusion; (2) rapidly diversifying student demographics; and (3) evidence that student success and retention were being impaired because UCSC's sustainability efforts did not sufficiently reflect these diverse backgrounds and lived experiences.

PoCSC raises awareness about the contributions that people of color have made to the environmental sustainability movement and fosters critical dialogue about environmental justice through events, speaker presentations, research on campus, student discussion spaces, social media awareness campaigns, courses, and workshops. This collective aims to recognize, highlight, and institutionalize the sustainability practices of communities of color, and aims to redefine sustainability to include diverse cultural approaches as well as a recognition of power dynamics, ethical considerations, and intersectional identities—what we call inclusive sustainability (Lu et al., 2018). PoCSC is the recipient of multiple awards: the 2016 UCSC Chancellor's Achievement Award for Diversity; 2017 California Higher Education Sustainability Conference's (CHESC) Best Practice Award in Sustainability Innovations; 2018 Indigenous Peoples Knowledge Community of the National Association of Student Personnel Administrators' (NASPA) Outstanding Student Program Award; 2020 CHESC Best Practice Award in Social Equity and Justice; and 2021 Finalist for the Racial Equity and Sustainability Collaborations Award, Association for the Advancement of Sustainability in Higher Education.

Over the years, PoCSC received many inquiries by faculty and staff at other universities to share our model, best practices, and research efforts towards building more equity-centered and inclusive environmental programming. This book enables us to do so. We bring together a diverse set of voices and areas of expertise (e.g., anti-colonial, anti-racist, community engaged, and critical feminist) that are largely lacking in the sustainability education literature. We look to these voices to disrupt normative notions of sustainability and to lead us into more critical visions of what campus-based sustainability can look like. This allows for issues of sustainability to be linked to a critical pedagogical approach that centers the ideas, practices, and demands by BIPOC scholars, students, and other community members in a more equitable fashion.

The Nexus of Environmental Sustainability, Social Justice, and Higher Education

The 1987 World Commission on Environment and Development published a report entitled, *Our Common Future* (commonly called the Brundtland Report after the Commission's Chairwoman and Norwegian Prime Minister Gro Harlem Brundtland), in which it provided the most oft-cited definition of sustainable development as "development that meets the needs of the present without compromising the ability of future generations to meet their own needs" (WCED, 1987: 43). The subsequent sentences in the report make clear that this entails "meeting the basic needs of all and extending to all the opportunity to fulfill their aspirations for a better life." Widespread and endemic poverty is not only "evil" and "no longer inevitable" but makes the world "prone to ecological and other catastrophes."

Our Common Future conveys the interconnections between social, environmental, and economic aspects of sustainability. As articulated by the Association for the Advancement of Sustainability in Higher Education (AASHE 2022), this sentiment is often conveyed as an alliterative three-legged stool or three-circle Venn Diagram of "economy, ecology, and equity" or "people, profits, and planet." The importance of the trifecta taken together and into account seems like an unalloyed good, a win-win-win. But the pillars can, and often are, in conflict, and when that happens the resolution invariably maintains dominant structures of power and privilege (Rose & Cachelin, 2018). Furthermore, as Redclift (1987, 1993) reminds us, sustainability and "its action-oriented variant" (Agyeman et al., 2002: 80), sustainable development, are fraught with confusion and contradiction. Such ambiguous and conflicting definitions not only make it possible for numerous political and social agendas to invoke sustainability, but "sustainable development has gained currency precisely because of the way it can be used to support these various agendas" (Redclift, 1993: 3).

Among the three legs of the stool, the social/societal/equity-oriented pillar tends to be the least developed[1] compared to biophysical environmental concerns and a progress-oriented approach that conflates "development" with "economic growth" (Vallance et al., 2011). "Development is read as synonymous with progress, and made more palatable because it is linked with 'natural' limits, expressed in the concept of sustainability" (Redclift, 1993: 7). This is perhaps unsurprising since the sustainability agenda was conceived by international committees, governmental and non-governmental networks,

[1] There is an area of scholarship about "social sustainability" seeking to remedy this gap; see for example Vallance et al., 2011; Dempsey et al., 2011; Eizenberg & Jabareen, 2017.

and think tanks (Ferreira, 2017; Agyeman et al., 2002), and reflects Western, post-Enlightenment, and Modernist thought and values (Redclift, 1993) that espouse themes of universalism, scientism, and technocentrism (Feinstein & Kirchgasler, 2015). As Redclift (1993: 19) elaborates:

> Sustainable development answers to problems initiated in the north, with our "global" development model. We have also understood sustainable development within a cultural context of our own—*our* view of nature, *our* problems with science and technology, *our* confidence in the benefits of economic growth…our capacity to find solutions is seriously reduced by our inability to recognize we are the prisoners of our history… which contributed to global environmental problems in the first place and made us poorly equipped to deal with them.

Ferreira (2017: 4) calls for the need to politicize sustainability, "to build a critical perspective *of* and *about* sustainability" that "scrutinize[s] the sources of degradation and exploitation… [to disrupt] the power relations that sustain oppressive structures." He frames this as an act of *conscientização* or cultivating critical consciousness and conscience (Freire, 1993). A critical sustainability, according to Ferreira, must counteract an ethos of dehumanization and exploitation situated in dominant power structures of colonialism, capitalism, white supremacy, heteropatriarchy, etc. and instead "engage with emancipatory understandings of power and power relations…and create theories and visions of sustainability that can lead to the development of more just, place-based cultures and social ecologies" (10). Framing sustainability as multifaceted, intersectional, multiscalar, systemic, and inclusive is necessary to avoid perpetuating dominant approaches that replicate current power dynamics and harm marginalized people and natural ecosystems. Sustainability is ontological, epistemological, axiological, and behavioral; thus, achieving a more sustainable society is not just about taking action, but being clear about the values undergirding those actions, the knowledges that inform such action, and the larger worldviews and theories of change that render those actions sensical.

Ferreira proposes a "critical sustainability studies framework of socio-ecological conscientization" (2017: 11) that borrows from critical Indigenous and ethnic studies, cultural studies, postcolonial theory, political ecology, and other fields with the aim of bringing together sustainability (with its focus on ecology and natural sciences) with critical theory. There are multiple facets that Ferreira articulates that resonate with our approach in this edited volume. First, hegemonic notions of sustainability—and related colonialist ideas such as wilderness (Cronon, 1996) and the nature/culture divide (Flikke, 2014; Shultis & Heffner, 2016; Norgaard, 2019; Álvarez & Coolsaet, 2020)—must be troubled, contested, and rearticulated "so that it can incorporate and critically engage with emancipatory understandings of power and power relations" (Ferreira, 2017: 9). Second, stop using the singular "sustainability," as though there is only one (i.e., Western) way to care for the planet; such discursive, ideological, cultural, and linguistic broadening opens the possibility for more heterogeneous, nuanced, and contextualized efforts toward place-based, justice-centered sustainabilities (Ferreira, 2017; Rose & Cachelin, 2018). Third, critical sustainabilities necessitate second-order change, in which the existing systems that perpetuate existential harm are questioned at their base, undermined, and subverted (Rose & Cachelin, 2018: 519). Such interrogations tend to be lacking in traditional, techno-scientific, and managerial approaches to sustainability, which have narrowly defined sustainability in scientific terms and positivistic definitions, sidestepping political analysis and rendering them largely devoid of the deeper social contexts in which sustainability is embedded.

These above tenets are a starting point toward delineating critical sustainability studies as a praxis for universities and other institutions of knowledge (co)production. Sustainability education is poised for a renewal, not only because of escalating concerns around climate change, but also due to demands to center issues of race, class, power, justice, and equity. As Heather McGhee, a Black climate activist, bestselling author and policy advocate, stated,

> It's essential to have anti-racism baked into the goals that even white-led organizations are pursuing because both political racism and environmental racism are drivers of our excess pollution and climate denialism...Success is measured by the improvement in the environmental and economic health of the people who have borne the brunt of our carbon economy (Sengupta, 2020).

Along with many other US universities, UCSC, which serves a majority student population of color, still grapples with its long history (and contemporary present) as a predominantly white, settler colonial institution situated on unceded territory. UCSC has long touted its renown as an institution committed to environmental stewardship and social justice, yet often falling short of bridging the two. As UCSC undertakes another campus-wide, Chancellor-led strategic planning process in AY 2022–2023 that strives for "climate change, sustainability and resilience" and "inclusive and thriving campus community" as two of five key goals, this is a tremendous opportunity for universities to convey and effectuate their values and commitments. Wals and Jickling (2002) suggest that the "squishiness" of the concept of sustainability allows institutions to hone in on what is most important to them in defining it. As they put it,

> The fact that 'sustainability' is a messy, ill-defined concept gives universities an opportunity to grapple with the concept and develop new ways of thinking about the concept. Sustainability provides colleges and universities an opportunity to confront their core values, their practices, their entrenched pedagogies, the way they program for student learning, the way they think about resources and allocate these resources and their relationships with the broader community (230).

This might be overly optimistic. Sehrsweeney (2020) argues that "Covid-19 is just the warm-up act for climate disaster. If universities handle climate change the way they've handled the pandemic, we're in big trouble." Universities—with their research capacities, public service orientation, political and economic capital—can and need to take drastic, decisive, and visionary action to avert catastrophic public health crises. Despite invoking a "climate-justice agenda", Sehrsweeney's suggestions for universities to meet this moment center on facilities and infrastructural changes toward carbon neutrality, from retrofitting buildings and clean energy generation to making campuses more bike friendly. While he correctly notes that the pandemic and climate change "disproportionately affect poor, non-white, and otherwise oppressed communities," and that increasingly corporatized universities use epidemiological and ecological crises to implement economic austerity that targets the most vulnerable for layoffs and furloughs, Sehrsweeney nonetheless calls for "action from the top down."

In contrast, calls for critical sustainabilities programming, education and collaborations are often led by students, staff, and community partners, especially those from historically underrepresented and marginalized groups, whose voices are provocations for the institution to think and act more expansively. This book amplifies some of these voices and bottom-up efforts toward a more critical approach to sustainability on campus, work that centers themes of power relations, (in)equity, (in)accessibility, and social (in)justice to study the interrelationships between humans, non-humans, and the environment. *As critical sustainabilities scholars, teachers, learners, and practitioners, we are committed to (1) disrupting normative narratives and epistemic hierarchies, centering underrepresented voices, and unsettling dominant understandings by pluralizing the discussion to bring in important knowledges; (2) elevating relationality and connection, between people and place, with each other and non-humans; (3) confronting settler colonialism and white supremacy in higher education; and (4) centering collective care.* We elaborate each of these points below.

1. Di Chiro and Rigell (2018) describe the formation of the Sustainable Serenity Collaborative, a partnership between Swarthmore College and the predominantly African American residents of north Philadelphia. In their efforts to "re-imagine and re-create sustainability," the authors emphasize the importance of "recognition and respect for diverse knowledge systems as central to the work (academic/theoretical knowledge, experiential knowledge, cultural knowledge, local knowledge)" (Di Chiro & Rigell, 2018: 80). Guided by a just sustainability approach, which centers broad accessibility of all to healthy environments and secure livelihoods, the Sustainable Serenity Collaborative intentionally utilized storytelling and community forums to articulate residents' hopes and dreams for their community. Their articulated desires for living-wage jobs, healthy food, and clean, safe streets led the collaborative to pursue efforts such as training for green-collar jobs (e.g., solar panel installation), energy-efficiency home retrofits, and a community garden. This is one example of disrupting epistemic hierarchies that often undermine potential collaborations between universities and local communities, and centering underrepresented voices and desire-based visions for what residents envision as sustainable.
2. The Sustainable Development Institute (SDI) at the College of Menominee Nation (CMN) is one of the first Indigenous-run research institutions and higher education programs in sustainability. The Menominee Nation, whose ancestral lands encompass what is now known as the Upper Midwest and Great Lakes region, maintains "cultural and spiritual practices [that] have served to motivate human responsibility within ecosystems" (Whyte et al., 2018: 152). According to the SDI website, "Menominee approach to sustainable development is an integration of

tribal wisdom, knowledge, values, understandings, and practices that views sustainability as a continual process by which Menominee affinity to place balances dimensions of community life." In their chapter about Indigenous lessons about sustainability, Whyte et al. (2018) discuss the cultural and spiritual practices that foster and enact human responsibility to promote ecosystem integrity, and how these processes changed and endured under conditions of settler colonial oppression. Indigenous ecologies supported flourishing arrangements of humans, nonhuman relatives (plants, animals, rocks), spiritual and inanimate entities, and landscapes—these thriving assemblages, the basis of futurity, are possible through moral relationships of reciprocity, trust, adaptation, and buffering. As evinced in the SDI, Indigenous planning and institutional governance offer a model for sustainability, one situated in the dynamic intersections between land and sovereignty; the natural environment (which includes humans); institutions (e.g., Tribal government, clan system, CMN); technology (e.g., from GIS to tools to make birch bark canoes); economics (i.e., systems and interactions that prioritize collective wellbeing); and human perception, activity, and behavior. Sustainable development is the "process of maintaining balance and reconciling the tensions within and among the six dimensions of sustainability" (Whyte et al., 2018: 168–169). Menominee worldviews exemplify the interconnectedness and relationality between human, non-human beings, and place.

3. The American academy is established through Indigenous dispossession and anti-Blackness, as institutions parasitized off enslaved labor and Indigenous Land. As far back as 1502 in the Americas, the Catholic church traded, bought, and sold African slaves in Atlantic markets as it assembled institutions in the New World with faculty to study and coerce the *indios* to abandon their eco-spiritual beliefs and non-European languages. The multi-sited project of missionization sought to civilize Indigenous peoples through worship of a singular white embodiment of God who could cast eternal damnation, justifying all non-whites' subjugation and naturalized inferiority (Fanon, 2007: 42; Rizzo, 2016: 27). New World academies, as sites of Eurocentric knowledge production, sought to replace Indigenous, earth-based cosmologies and spiritualities.

On the cusp of the Emancipation Proclamation that would abolish chattel slavery, 1862 was a pivotal year for the implementation of colonial federal policy to expropriate Indigenous lands. President Abraham Lincoln first signed the Homestead Act, creating 1.5 million homesteads for settler-farmers to establish themselves on more than 300 million acres of unceded Indigenous territory west of the Mississippi River. He then signed the Morrill Act, which lined state coffers with land-handouts for a national-scale production of colleges and universities. By the 1900s, the federal government had redistributed 10.7 million acres of Indigenous land to 52 land-grant universities, amassing $22.8 million of endowment funds in principal and unsold landholdings (Lee & Ahtone, 2020). The University of California, to which UCSC belongs, leveraged these massive land swaths for real estate brokerage and financing that covered large portions of their annual expenses.

In her excellent scholarship revealing the racist and settler colonial roots of the University of California system, Sabati (2019) traces the origins of the UC, the preeminent system of public higher education in the nation, focusing on the latter half the nineteenth century in which "logics of white supremacy were systematically institutionalized throughout the state [of CA] and supported its transition to modern capitalism" (66). Sabati describes various examples of material and epistemological forms of racial and colonial violence, from brutality inflicted against Indigenous peoples during Spanish missionization, Mexican period, and Gold Rush and White settler violence, to multiple legislative efforts to limit the movement of Blacks to California, to the exploitation of Chinese immigrant labor. Notions of who were considered "good citizens" or even human were racially constituted, and "the moral order became co-instantiated with the economic, legal and political order" (Sabati, 2019: 79). This order, which elevated reason as a key attribute of humanness, could be determined by a university system with epistemological authority, one that also inculcated ecumenical Christian virtues, promoted democratic values and progress, and produced "men of science and letters," all lofty goals that served to erase settler colonial violence, enact "colonial unknowing", and expand the US imperial project. Sabati describes a gathering of the Trustees of the first public university of California in the late 1860s, who extolled the work of Bishop of Cloyne George Berkeley, a slave-owner who actively worked to evangelize Indigenous people, as a man whose "efforts to spread Western thought and Christian values through the auspices of education captured the spirit of the institution they hope to build" (2019: 61).

Sabati argues that "Universities actively uphold and exacerbate contemporary contexts of racial stratification and settler colonialism" (2019: 68). We witness examples of this at UCSC, from the UCSC Astronomy department's ongoing fiscal and academic sponsorship of the "state-of-the-art" TMT telescope that would desecrate Indigenous Hawaiians' sacred site of worship on Mauna Kea's summit, to the fact that only in the summer of 2019 did the university agree to remove the Santa Cruz mission bell from its campus, which had served as a painful reminder of the genocide perpetrated against the ancestors of the Amah Mutsun Tribal Band and many other California tribes for more than two centuries (Hanna, 2019). In spring of 2017, more than 100 students joined members of the Afrikan Black Student Alliance to occupy Kerr Hall, the administrative building at UCSC. Their demands included a guarantee of four-year on-campus housing for all African, Black, and Caribbean students in RPAATH (the Rosa Parks African American Themed House); a repainting of that house in Red, Black, and Green colors of the Pan-Africanist Liberation movement; the creation of a Black Studies department on campus; and mandatory diversity competency training for all incoming students (Afrikan Black Coalition, 2017). In June 2020, after the murder of George Floyd, the UCSC Black Student Union issued a set of demands titled "The Unrelenting Anti-Blackness of 2020." After a year of inaction by the campus, these demands were revisited in May 2021 (Glover, 2021).

4. The Black Student Union's 2021 demands include: the hiring of more African, Black, and Caribbean (ABC)-identified and queer and trans-friendly psychologists at UCSC's Counseling and Psychological Services office; more ABC-themed housing for low-income and historically disadvantaged ABC students; and improvements in the manner in which incidents of hate, discrimination, prejudice, and bias are dealt with on campus. The BSU is asking for the campus to support their physical and emotional well-being and safety, a form (to allude to the vignette at the beginning of this chapter) of communal/individual sustainability. UCSC "fails to provide the resources necessary to sustain ourselves—housing, food, and access to transportation… When people have the resources they need to sustain themselves, we create sustainable communities" (Blueprint for a Sustainable Campus, 2019–2021: 24). Integrating Black radical thought, feminist, and critical scholarship, Ranganathan and Bratman (2021: 121) recognize that "struggle must be understood as intersectional—not only in the sense of overlapping identities…but also in the sense of materialities and lived experiences" that understand the "environment" more capaciously to include basic needs and everyday threats.

Critical sustainabilities entail a commitment to collectivity, care, and their integration as collective care. Collectivity is not often a concept at the fore in higher education cultures and spaces, which tend to prioritize self-interest rather than collective well-being. Faculty are rewarded for their research (over teaching, mentoring, and service) and incentivized to prioritize their own professional stature and reputation through a focus on solo authored works, invitations to give keynote addresses, their citational index, etc. "The neoliberal model places the emphasis on economic metrics whereby students are regarded as consumers, and faculty as mobilizers of particular forms of commodifiable knowledge" (Gardiner, 2020: 6). The educational system is "structured by output, defined and operationalized as the production of things for the advancement of the institution in a competitive market" (Wilson & Richardson, 2020: 82; Motta & Bennett, 2018). Gardiner (2020) draws on the work of Hannah Arendt to argue that individualized and hierarchical forms of leadership, common in university settings, is oppositional to collective action and undermines human flourishing. The tendency to elevate individual success over collective well-being has been found to undercut student performance, especially among underrepresented groups. Covarrubias et al. (2016) undertook a study about student academic performance that contrasted the independent/autonomous model of the self, predominantly fostered in White, middle-class contexts, with the interdependent model of the self, fostered in more collectivist Asian, Latinx, Native American communities, and working-class contexts. They found that for Latinx students, culturally relevant affirmations that focused on the family improved academic performance as compared with self-affirmation, whereas for White students there was no difference. The predominant independent-focused cultural norms of institutions of higher education are thus likely to mismatch with the values of students from interdependent backgrounds (e.g., working class, students of color).

Sustaining students at the university thus necessitates challenging a culture of hierarchy, individualism, and competition and supporting a culture of collectivity, care, and solidarity. Care is a "response to the oppressive conditions found in the world and in the school spaces we work in…it creates the conditions for well-being, conditions that institutions of higher education (structurally) have yet to create and maintain" (Wilson & Richardson, 2020: 80). Motta and Bennett (2018) frame care within the context of higher education as: first, a recognition and embracing of the whole student, rejecting deficit and damage-centered discourses and approaches; second, as dialogic relationality within an onto-epistemological commitment of mutuality between teachers and students; and third, as affective and embodied praxis. Ginwright (2018) takes this conceptualization a step further, noting that "well-being comes from participating in transforming the root causes of the harm within institutions" and that "trauma and healing are experienced collectively." We draw on his model of Healing-Centered Engagement to describe collective care as a component of critical sustainabilities in higher education. Well-being is a function of our environments and our ability to have efficacy within the spaces where we live, work, and play. Through collective, culturally resonant, and asset-driven efforts to build awareness of and address oppression and harm, we gain a sense of purpose, power, and control.

Organization of the Edited Volume

Critical Campus Sustainabilities is organized into three parts. Part I, entitled, "Hearing and Addressing Students' Needs," speaks to the lived experiences and concerns of students, both inside and outside of the classroom, that reflect some of the limitations of traditional ideas about sustainability and how to ameliorate them.

In "Inclusive Sustainability: The emergence and vision of PoCSC," authors Rebecca Hernandez, Elida Erickson, and Adriana Renteria discuss the history and foundations of the People of Color Sustainability Collective. The organization was founded as a way to provide not only a space for students of color to partake in environmental conversations, but also as a way to infiltrate the boundaries of environmental conversations, notions of sustainability, and campus agendas. PoCSC's goal is to uplift students of color and their relations to the environment, pushing the conversation to not only include students of color but also to change the normative ethos of environmental sustainability. This chapter: (1) discusses the history of UCSC's campus sustainability program as largely consistent with other institutions; (2) explains the emergence of PoCSC as an effort to foster a more "inclusive sustainability," a concept coined by the initiative; and (3) highlights the intentionality of PoCSC's design, namely housing it within thc Ethnic Resource Centers and specifically the American Indian Resource Center, which centered issues of settler colonialism and essentialization of Native Americans within mainstream environmental sustainability.

The next chapter, entitled, "Student Voices on Environmental Spaces and Experiences in Higher Education," Kimie Dare and Riri Shibata facilitate and comment on a roundtable discussion of various Black, Indigenous, and people of color participants, who, like them, are alumni in the Department of Environmental Studies at UCSC. They note the contradiction between UCSC's liberal, progressive ideals and the department's renowned reputation and the roundtable participants' feelings of exclusion, isolation, and disappointment. These sentiments are part of a larger, consistent pattern among BIPOC students in environmental fields. The participants do not see themselves or their interests represented or welcomed in the department. Instead, they experience pervasive yet invisible slights that signal their otherness, like being the sole person of color in an entire class or not relating to comments made by peers or professors. They voice their frustrations while identifying other campus resources and spaces—like co-curricular programs like the PoCSC and courses in the Critical Race and Ethnic Studies department—where they feel empowered, heard, and supported.

In the final chapter of Part I, "Teaching Critical Sustainability Studies: Towards a Relational Pedagogy," Emily Murai discusses the ways a relational pedagogical approach can enhance core concepts in critical sustainability studies and redress the shortcomings of curricula that has been shaped by mainstream environmentalism's emphasis on ecology, conservation, and preservation. Courses that emphasize both relationality and a critical lens have the potential to attract a diverse range of students whose interests, identities, and relationships to the environment are better understood through the framework of social justice. In the chapter, Murai discusses specific course design strategies that enable students to develop both theoretical knowledge and practical skills to work toward the aims of critical sustainability. Through this, she shows how both course content and classroom practices can enhance each other to enable students to work toward more inclusive environmentalism beyond the classroom.

Part II, "UCSC Students' Sustainability Perceptions, Understandings and Values," reports key findings from three campus-wide, undergraduate surveys conducted by the People of Color Sustainability Collective in 2016, 2019, and 2022, a period that spans the election of Donald Trump, the Covid-19 pandemic, racial justice reckoning after the murder of George Floyd, the election of Joe Biden, and an onslaught of multiple destructive natural disasters across the US and beyond. With over 8,000 undergraduate respondents total, this is one of the largest datasets of its kind.

In "Student Understandings of Sustainability," authors Flora Lu, Tashina Vavuris, Randy Uang, Cassandra Wood, and Anna Sher explore the wide array of understandings attributed by undergraduates to the term "sustainability". On the one hand, such vagueness offers heuristic capacity and educational potential (Wals & Jickling, 2002); on the other hand, ambiguity enables the ability of powerful actors to reduce the concept to particular framings that support dominance, control, and obscuring of alternatives. In this chapter, we survey UCSC undergraduates in 2019 and 2022, asking them to rank five definitions of sustainability in descending order of importance. We explore how definitional rankings changed over time and differed by student characteristics (i.e., race, gender, first-generation versus continuing-generation status, and disciplinary major). It is clear that there is a plurality of definitions of sustainability among our students, which means that no one definition tends to dominate; therefore, we need to operationalize and be explicit about meanings lest we talk past each other. The point is not to settle on one general, universal definition of sustainability, but instead grapple with how various interpretations reflect power dynamics, agendas, and worldviews that must be recognized and reconciled. These complex, contested, and concurrent framings of sustainability constitute a useful and necessary space for critically interrogating these interpretations and the interests, goals, trade-offs, and visions for the futures they represent.

Next, Flora Lu, Randy Uang, and Anna Sher explore the contradictions of the "environmental belief paradox," which refers to the prevalent, widely held, and erroneous notion that white people are the most environmentally conscious. In other words, it is "a tendency to misperceive groups that are among the most environmentally concerned and most vulnerable to a wide range of environmental impacts as least concerned about the environment" (Pearson et al., 2018: 12430). This pervasive notion is held by both higher-income whites, who believe that their demographic is the most concerned about sustainability, and low-income people of color, who similarly underestimate the environmental attitudes of the groups to which they belong. In light of evidence that some UCSC students discount the environmental concerns of students of color (consistent with the environmental belief paradox), we share survey data from 2019 and 2022 from undergraduates who self-reported their levels of concern about various environmental issues—from buildings and transportation, to protecting water/land/ecosystems, to biophysical and cultural sustenance—and we explore how environmental concerns vary not just by race and ethnicity, but also gender, first-generation status, and major.

The following chapter by Tashina Vavuris, Cassandra Wood, Flora Lu, Randy Uang, and Anna Sher is entitled, "Environmental Sustainability and Epidemiological Struggle: Student Experiences of COVID-19." The COVID-19 pandemic revealed glaring disparities in vulnerability and precarity in our society, from those who could afford to limit their exposure to the virus and maximize the quality of care if infected, to those most at risk to sickness and death. Environmental conditions, from air contamination to access to greenspace, significantly mediated how people experienced the pandemic. This chapter reports survey results from 2022 that asked undergraduate students about the environmental health threats impacting their home communities and the related health issues (e.g., asthma, lead poisoning, etc.) they and their families experience. Over half of the respondents said that their communities suffer from air pollution; 44% reported housing insecurity; and about a third named lack of greenspace, food insecurity, and water contamination as problems. Environmental health problems are unevenly felt. For example, 35% of students of color reported that someone in their family has a health condition related to environmental problems, compared to 31% of white students; 40% of first-generation students answered affirmatively, compared to 30% of continuing-education students. All survey respondents reported that environmental health threats worsened their experience of the

COVID-19. Food and housing insecurity, inadequate sanitation, and lack of greenspace were the top issues reported, and we found differences by race and first-generation status.

Finally, the last chapter in Part II, by Flora Lu, Emily Murai, Serena Campbell, and Hillary Angelo, is "Critical Environmentalisms: Overcoming Institutional Obstacles to Meet Students' Demands for Sustainability Curricula and Action." This chapter marshals longitudinal data from the three PoCSC undergraduate surveys to explore how students' concern about a dozen different environmental issues has changed from 2016 to 2022. The results consistently show that the percentages of students who express that environmental problems are "important" or "very important" increase dramatically. In 2016, for example, only one issue had more than 90% of respondents express concern, but by 2022, two-thirds of the issues met this bar. The results are largely consistent between students of color and white students. We also found that while in 2016 students felt that the campus cared more on a few issues but roughly the same as they did, by 2019 the opposite was true, with students reporting that the campus placed less importance than they did, with a gap anywhere between 5% to over 20%. By 2022, these disparities were even worse. The chapter then explores what sustainability-related topics students are learning about, how they assess changes in their understanding of environmental concerns, and what they want more curricular attention on, a result that differs between white and non-white students. We then draw on another research effort with surveys and interviews with faculty across campus to explore how the campus can promote more critical environmental research, teaching, and collaboration.

Part III, "Community-Engaged Critical Sustainabilities," addresses how universities can engage students, staff, faculty, and community members in sustainability-oriented research, curricular and co-curricular efforts, outreach, education, and collaboration. This section moves beyond UCSC to incorporate projects and research being done that centers community and non-traditional perspectives in different capacities and pushes the boundaries of transformative pedagogical spaces.

In the first chapter of this section, Linnea Beckett and Michelle Hernandez Romero describe "Developing a Praxis of Loving Relations: Lessons from a Community-University Partnership that Centers Undergraduate Research and Learning." The chapter highlights the work of the Apprenticeship for Community Engaged Research or (H)ACER program housed at John R. Lewis College at UCSC. As a link between community-engaged scholarship and undergraduate education, (H)ACER offers a variety of programming from one-day workshops or volunteer workdays, to full-credit courses and year-long, credit-bearing research apprenticeships at various field sites to train students in anti-racist and anti-colonial community engagement. The chapter draws from Hernandez Romero's experiences as a research apprentice in a nearby community garden to show the potential of justice-centered co-curricular programming within higher education. Hernandez Romero, a first-generation DACA college student, describes the "loving relations" (Lugones, 2003) she built with the gardeners who hail from the same community she is from in Mexico, and how these relations in turn transformed and supported her academic journey at UCSC.

Michelle Hernandez Romero, Yolanda Perez, and Linnea Beckett further extend discussion of the community garden introduced in the previous chapter in "*Plantando Amor y Cultivando Unidad:* A Story of Building and Sustaining an Immigrant-Led Community Garden in Partnership with a University." This chapter illuminates the development of the Calabasas Community Garden (CCG) in Watsonville, CA, a rural agricultural town, and the community-building processes that took place to create what is now seen as a space for community unity. In addition, the CCG is sustained through partnerships between UCSC, Calabasas Elementary School, and Watsonville community members, creating a unique collaboration and multifaceted space in which different members from the different partners come together to promote food justice and community empowerment and care. Each co-author is a member of one organization and sheds light on what the garden means to them and how they have managed to work together through various challenges, like bridging cultural gaps between members of the university with community gardeners, from their positionalities and perspectives.

In their chapter, "Environmental Justice Youth leadership in Salinas Valley, CA," Karen Crespo Triveño and Agustin Angel Bernabe describe their efforts with Latinx youth to create an environmental justice youth leadership academy (EJYLA) in an agricultural region where low-income, Latinx communities experience disproportionate levels of environmental harm, from pesticide drift to siting of waste facilities. Not only is the EJYLA a space for students to share their experiences from environmental harm, but it is also a way for dreams, hopes, and desires to be re-imagined around the topic of environmental sustainability. This chapter situates the academy within the scholarship of critical environmental justice, youth activism, and activist histories in the Salinas Valley and discusses its development and implementation in the summer of 2021. The authors' reflections show the social justice implications that co-constructed youth education programs can have on environmental justice education, community health and well-being, and participatory epistemologies.

In the Conclusion, Christopher Lang, Emily Murai and Flora Lu summarize the context of the book and illuminate future pathways for sustainability education. In particular, we advocate for a broad, inclusive approach to sustainability within higher education, one that understands the interconnection between seemingly disparate social justice issues and emphasizes collective care and well-being as a means to best serve the interests of students and the next generation of environmental activists. For inspiration, we look at the #COLA4ALL wildcat strike that took place at UCSC in late 2019 through early 2020 as a way to formulate a critical sustainabilities movement in higher education. In addition to its primary call for a cost-of-living adjustment, #COLA4ALL articulated multiple demands to improve the living and working conditions of the university community and work toward racial justice and long-term economic sustenance in public higher education. Post-COVID, the movement resurfaced in the 2022 labor strikes that were led by four graduate student worker unions across all ten University of California campuses. After six weeks of picketing, the unions ratified labor contracts despite ongoing controversies and internal disputes. As a result, the #COLA4ALL movement continues to persist in calling for living wages and affordable housing for UC knowledge workers among other demands for justice. Similarly, we support students in their continued advocacy for an overlapping, multi-interest, and inclusive approach to sustainability as the surest path toward environmental longevity.

References

Afrikan Black Coalition. (2017). Black students at UC Santa Cruz protest hostile campus. *Santa Cruz IMC*. Retrieved September 15, 2022, from https://www.indybay.org/newsitems/2017/05/03/18799121.php

Agyeman, J., Bullard, R. D., & Evans, B. (2002). Exploring the nexus: Bringing together sustainability, environmental justice and equity. *Space and Polity, 6*(1), 77–90. https://doi.org/10.1080/13562570220137907

Álvarez, L., & Coolsaet, B. (2020). Decolonizing environmental justice studies: A Latin American perspective. *Capitalism Nature Socialism, 31*(2): 50–69. https://doi.org/10.1080/10455752.2018.1558272

Association for the Advancement of Sustainability in Higher Education. (2022). What is sustainability? Accessed January 23, 2023, from https://stars.aashe.org/resources-support/help-center/the-basics/what-is-sustainability/

Cachelin, A., & Rose, J. (2018). Guiding questions for critical sustainabilities. *Journal of Environmental Studies and Sciences, 9*, 570–572. https://doi.org/10.1007/s13412-018-0508-3

College of Menominee Nation. (n.d.). About SDI. Accessed 23 Jan 2023, from: https://www.menominee.edu/sustainable-development-institute/about-sdi

Covarrubias, R., Herrman, S. D., & Fryberg, S. A. (2016). Affirming the interdependent self: Implications for Latino student performance. *Basic and Applied Social Psychology, 38*(1), 47–57. https://doi.org/10.1080/01973533.2015.1129609

Cronon, W. (1996). The trouble with wilderness: Or, getting back to the wrong nature. *Environmental History, 1*(1), 7–28.

Dempsey, N., Bramley, G., Power, S., & Brown, C. (2011). The social dimension of sustainable development: Defining urban social sustainability. *Sustainable Development, 19*(5), 289–300. https://doi.org/10.1002/sd.417

Di Chiro, G., & Rigell, L. (2018). Situating sustainability against displacement: Building campus-community collaboratives for environmental justice from the ground up. In J. Sze (Ed.), *Sustainability: Approaches to environmental justice and social power* (pp.76–101). NYU Press.

Eizenberg, E., & Jabareen, Y. (2017). Social sustainability: A new conceptual framework. *Sustainability, 9*(1), 68. https://doi.org/10.3390/su9010068

Fanon, F. (2007). *The wretched of the earth*. Grove/Atlantic, Inc.

Feinstein, N. W., & Kirchgasler, K. L. (2015). Sustainability in science education? How the next generation science standards approach sustainability and why it matters. *Science Education, 99*(1), 121–144. https://doi.org/10.1002/sce.21137

Ferreira, F. (2017). Critical sustainability studies: A holistic and visionary conception of socioecological conscientization. *Journal of Sustainability Education, 13*, 1–22.

Flikke, R. (2014). On the fractured, fragmented and disrupted landscapes of conservation. *Forum for Development Studies 41*(2), 173–182. https://doi.org/10.1080/08039410.2014.918759

Freire, P. (1993). *Pedagogy of the oppressed*. Continuum Books.

Gardiner, R. (2020). Caring leadership as collective responsibility. In L. Tomkins (Ed.), *Paradoxes of leadership and care: Critical and philosophical reflections* (pp. 175–186). Edward Elgar.

Ginwright, S. (2018). The future of healing: Shifting from trauma informed care to healing centered engagement. Retrieved May 31, 2018, from https://ginwright.medium.com/the-future-of-healing-shifting-from-trauma-informed-care-to-healing-centered-engagement-634f557ce69c

Glover, S. (2021, May 20). After a year of inaction, students of BSU revisit demands. *City on a Hill Press*. https://www.cityonahillpress.com/2021/05/20/after-a-year-of-inaction-students-of-bsu-revisit-demands/

Hanna, J. (2019, June 19). A California university removes a mission bell from campus, concerned it's seen as a racist symbol. CNN. Retrieved September 13, 2022, from https://www.cnn.com/2019/06/22/us/university-california-santa-cruz-mission-bell

Lee, R., & Ahtone, T. (2020, March 30). Land-grab universities. *High Country News*. https://www.hcn.org/issues/52.4/indigenous-affairs-education-land-grab-universities

Lu, F., Rosser, R. H., Renteria, A., Kim, N., Erickson, E., Sher, A., & O'Connor, L. (2018). Inclusive Sustainability: Environmental Justice in Higher Education. In Leal Filho, W., Marans, R. W. & J. Callewaert (Eds.), Handbook of Sustainability and Social Science Research (pp. 63–81). Springer

Lugones, M. (2003). Pilgrimages/peregrinajes: Theorizing coalition against multiple oppressions. Rowman & Littlefield Publishers

Motta, S. C., & Bennett, A. (2018). Pedagogies of care, care-full epistemological practice and 'other' caring subjectivities in enabling education. *Teaching in Higher Education, 23*(5), 631–646. https://doi.org/10.1080/13562517.2018.1465911

National Academies of Sciences, Engineering, and Medicine. (2020). *Strengthening sustainability programs and curricula at the undergraduate and graduate levels*. The National Academies Press. https://doi.org/10.17226/25821

Norgaard, K. (2019). Nature, culture and the production of race. *Section Culture: Newsletter of the ASA Culture, Section, 31*(1). Retrieved January 23, 2022, from https://asaculture-section.org/newsletters/

Pearson, A. R., Schuldt, J. P., Romero-Canyas, R., Ballew, M. T., & Larson-Konar, D. (2018). Diverse segments of the US public underestimate the environmental concerns of minority and low-income Americans. *Proceedings of the National Academy of Sciences, 115*(49), 12429–12434

Ranganathan, M., & Bratman, E. (2021). From urban resilience to abolitionist climate justice in Washington, DC. *Antipode, 53*(1), 115–137. https://doi.org/10.1111/anti.12555

Redclift, M. (1987). *Sustainable development: Exploring the contradictions*. Routledge.

Redclift, M. (1993). Sustainable development: Needs, values and rights. *Environmental Values, 2*(1), 3–20.

Rizzo, M. A. (2016). *No Somos animales: Indigenous survival and perseverance in 19th century Santa Cruz, California* (Doctoral dissertation). https://escholarship.org/uc/item/72n1q0vz

Rose, J., & Cachelin, A. (2018). Critical sustainability: Incorporating critical theories into contested sustainabilities. *Journal of Environmental Studies and Sciences, 8*, 518–525. https://doi.org/10.1007/s13412-018-0502-9

Sabati, S. (2019). *Ethical elisions: Unsettling the racial-colonial entanglements of U.S. Higher Education* (Doctoral Dissertation). https://escholarship.org/uc/item/16g4q7cf

Sehrsweeney, M. (2020). Covid-19 is just the warm-up act for climate disaster. *Chronicle for Higher Education*. Retrieved October 8, 2020, from https://www.chronicle.com/article/covid-19-is-just-the-warm-up-act-for-climate-disaster?

Sengupta, S. (2020, June 3). Black environmentalists talk about climate and anti-racism. *The New York Times*. https://www.nytimes.com/2020/06/03/climate/black-environmentalists-talk-about-climate-and-anti-racism.html?searchResultPosition=1

Shultis, J., & Heffner, S. (2016). Hegemonic and emerging concepts of conservation: A critical examination of barriers to incorporating Indigenous perspectives in protected area conservation policies and practice. *Journal of Sustainable Tourism, 24*(8–9), 1227–1242. https://doi.org/10.1080/09669582.2016.1158827

UCSC Enviroslug. (2019–2021). Blueprint for a sustainable campus. Accessed 23 Jan 2023, from http://enviroslug-sec.org/blueprint

Vallance, S., Harvey C. P., & Dixon, J. E. (2011). What is social sustainability? A clarification of concepts. *Geoforum, 42*, 342–348.

Wals, A., & Jickling, B. (2002). Sustainability in higher education: From doublethink and newspeak to critical thinking and meaningful learning. *International Journal of Sustainability in Higher Education, 3*(3), 221–232.

Whyte, K., Caldwell, C., & Schaefer, M. (2018). Indigenous lessons about sustainability are not just for "all humanity.". Sustainability: Approaches to environmental justice and social power, 149.

Wilson, A., & Richardson, W. (2020). All I want to say is that they don't really care about us: Creating and maintaining healing-centered collective care in hostile times. *Occasional Paper Series* (43). https://doi.org/10.58295/2375-3668.1342

World Commission on Environment and Development. (1987). *Our common future*. Oxford University Press.

Part I

Hearing and Addressing Students' Needs

1 Inclusive Sustainability: The Emergence and Vision of PoCSC

Elida Erickson, Rebecca S. Hernandez, and Adriana Renteria

Abstract

The People of Color Sustainability Collective (PoCSC) was founded as a way to provide not only a space for students of color to participate in environmental conversations, but also as a way to infiltrate dominant/mainstream environmental discourses, notions of sustainability, and campus agendas. PoCSC's goal is to uplift students of color and their relations to the environment, pushing the conversation to not only include students of color, but also to change the normative ethos of environmental sustainability. This chapter will: (1) discuss the history of UCSC's campus sustainability program as largely consistent with other institutions; (2) explain the emergence of PoCSC as an effort to foster a more "inclusive sustainability," a concept coined by the initiative; and (3) describe the intentionality of PoCSC's design, namely housing it within the Ethnic Resource Centers and specifically the American Indian Resource Center, which centered issues of settler colonialism and essentialization of Native Americans within mainstream environmental sustainability.

Keywords

Inclusive sustainability · Mainstream environmentalism · Black, Indigenous, and People of Color (BIPOC) · Essentialization · Tokenization · Higher education · University of California

Introduction

In this chapter, we discuss the climate of environmental sustainability at the University of California, Santa Cruz (UCSC) from the perspective of three different individuals involved in the formation of the People of Color Sustainability Collective (PoCSC), who worked with students to bridge sustainability with diversity, equity and inclusion in ways that were innovative and unprecedented for the campus. The authors—a Director of the Sustainability Office, then-Director of the American Indian Resource Center, and a student leader turned PoCSC Program Coordinator at the time—each share their respective pathways to this important work over several years. Together with Dr. Flora Lu, a Professor of Environmental Studies, we built upon our complementary areas of expertise and shared commitment to providing students a welcoming,

E. Erickson (✉) · R. S. Hernandez
University of California, Santa Cruz, CA, USA
e-mail: ekericks@ucsc.edu

A. Renteria
California State Water Resources Control Board, Sacramento, CA, USA

F. Lu, E. Murai (eds.), *Critical Campus Sustainabilities*, Sustainable Development Goals Series,
https://doi.org/10.1007/978-3-031-30929-8_1

uplifting, and transformative undergraduate experience. Though higher educational institutions encourage partnerships between faculty and staff with expertise in student affairs, the necessity of these relationships has gained increased attention since the COVID-19 pandemic as a means to create multi-layered support networks for students (Skallerup & Fisher, 2021). Both our cross-campus collaboration and the opportunity to reflect on our work is rare and we believe that is the case with many colleagues who are committed to this type of effort. Our hope is that this chapter will provide both insights and practical advice to other practitioners who wish to create bridges between social justice and environmentalism and collaborations across disparate units at the university.

PoCSC was founded to provide not only a space for students of color to partake in environmental conversations, but also as a way to challenge dominant environmental conversations, notions of sustainability, and campus agendas. This initiative was several decades in the making, shaped by Santa Cruz culture, changing UCSC demographics, the challenges of mainstream sustainability and environmentalism, and growing student discontent. PoCSC's goal is to uplift students of color and their relations to the environment, expanding the conversation to not only include students of color, but also to change the normative ethos of environmental sustainability. Student activism, student lived experiences, and student success were core drivers in the evolution of the concept of "inclusive sustainability," a term coined by PoCSC.

This chapter will discuss the history of UCSC's campus sustainability program as largely operational and explain the emergence of PoCSC as an effort to foster a more inclusive sustainability. We will discuss the intentionality of PoCSC's design, namely housing it within the Ethnic Resource Centers and specifically the American Indian Resource Center, which centered issues of settler colonialism and essentialization of Native Americans within mainstream environmental sustainability, as discussed later in this chapter. We will share lessons learned and the best practices we have co-created. We hope that our reflections encourage campus leaders to reflect on their own roles in supporting student retention through inclusive sustainability programming.

Elida Erickson

By the time I joined the UCSC Sustainability Office in 2011, the campus had a long-standing reputation for advancing environmentalism through its academic curriculum, co-curricular programs, and operational initiatives. Far before the term "sustainability" became standard within academia by the early 2000s, the institution had a well-established and world-renowned environmental studies department, a 30-acre organic farm that trained apprentices from around the globe, and several innovative student-led organizations, including the Student Environmental Center.

Environmentalism was so ingrained within the campus culture that, in 2003, undergraduate students voted to enact a new student fee measure, and in 2006, undergraduate and graduate students voted for additional fees. These fees were committed to advancing environmental projects and initiatives across the campus, resulting in the formation of the Campus Sustainability Council and Carbon Fund student governing bodies that jointly manage over $500,000 in funds annually. The Education for Sustainable Living Program, launched in 2003, offers opportunities for undergraduate students to gain actual teaching experience through student-led courses on a broad range of sustainability topic areas. Additionally, the university is made up of ten distinct residential colleges focusing on different thematic areas. In 1972, College Eight was established with the theme "Environment & Society," and was notably re-named Rachel Carson College in 2016.

What unique characteristics about UCSC have drawn the campus so strongly towards environmental ethics? Many factors are likely at play, but most notable is its exceptional geographic location and topography. Nestled on the California coast at the base of the Santa Cruz mountains, the campus sits on rolling hillsides amongst coastal redwood trees and offers sweeping vistas of the internationally-renowned Monterey Bay Marine Sanctuary. One would be hard-pressed to visit the campus without contemplating the natural environment around them.

Another notable characteristic of UCSC is a strong history of student activism and advocacy. In 2013, the campus' University Relations referenced this legacy, creating the slogan "UCSC: The original authority on questioning authority" (University of California, Santa Cruz Public Affairs, 2013). UCSC is one of ten residential campuses within the University of California (UC) system, which has a strong history of student activism across many of its campuses. Students made history at UC Berkeley and other campuses when they protested against the Vietnam War in the 1960s. In the 1990s, student activism was instrumental in pushing the UC system to divest financial resources from entities associated with apartheid regimes in South Africa.

By the early 2000s, students were growing up in an era where people were becoming increasingly aware of climate change and were starting to accept it as scientific fact. Students in environmental disciplines were learning about the need to increase renewable energy resources and build greener buildings that reduce water use and conserve energy. However, these imperatives did not include a focus on examining the physical structures and buildings of the campuses themselves, a disconnect that was observable both at more urban campuses such as UC Berkeley, as well as campuses in peri-urban areas like UCSC. Many students began to ask themselves why.

Students organized once again to advocate for the UC system to take a serious look at the operational impact of its own campuses on the environment. They urged the UC Office of President to implement a policy that would compel campuses across the UC system to increase sustainability and start serving as institutional models to address one of the most pressing challenges of our times. In a speech in 2007, UC President Dynes' summarized the sentiment well when reflecting back on this time period: "We taught [sustainability] to students, and a remarkable thing happened. The students bought it, understood it, and were sufficiently energized by it that they came back to us and said, 'Well, if you believe what you teach, why aren't you practicing what you teach?'… Students came back and said, 'You've got us committed, now we want you committed.' And we went through a full change of heart and belief."

In 2004, the UC implemented the Clean Energy and Green Building policy, the first sustainability policy commitment by any university system. The policy would evolve and expand over time to include other key sustainability-related issues: climate protection and sustainable transportation in 2006; recycling, waste management, environmentally preferable purchasing in 2007; sustainable food service in 2009; and sustainable water systems in 2013 (University of California, 2021). In alignment with traditional environmental approaches, each of these topic areas was highly focused on goals to improve the operational aspects of UC institutions. Key elements of the policy focused on conserving resources where possible, reducing the environmental footprint of the resources that are used, tracking metrics, and implementing technological solutions to problems. Each of the specific topic areas became sections within a broader UC Sustainable Practices Policy, which began to serve as a guidepost to individual campuses.

While the UC system continued to advance sustainability policy goals, staff members at each of the campuses became actively involved by serving on committees to help inform the development of the UC Sustainable Practices Policy. Staff and faculty were also forming committees and groups within the individual campuses to focus on their unique sustainability needs and implement institutional structures to improve their environmental performance. At the behest of active student leaders, recommendations from staff and faculty, and the new UC policy, administrators at the individual campuses started forming their own Sustainability Offices as an official function within the university's operations.

UCSC's Sustainability Office was established in the mid-2000s with one professional staff member and a small handful of student staff. It quickly grew to four staff members and nearly 20 student staff by 2012 to become the most robustly staffed Sustainability Office across the UC system. By the time I took on the director role in 2015, we had five professional staff members and over 30 student staff, and our office had been moved up to report directly to the Vice Chancellor for Business and Administrative Services. This model was a testament to UCSC's strong commitment to sustainability. Even the Chancellor's vision statement for the campus included a mission to bolster "the values of social and environmental sustainability" (University of California, Santa Cruz, 2014).

As the UC systemwide policy continued to evolve, the UCSC Sustainability Office conducted a baseline assessment of the campus' environmental programs and operational performance in 2007, and developed its first ever three-year Campus Sustainability Plan in 2010. The 2010–2013 and 2013–2016 plans set goals in several topic areas including: Buildings, Food, Procurement & Purchasing, Recycling & Waste Reduction, Energy, Land & Habitat, Water, and Transportation. Just as with the system UC policy, goals were largely operational in focus with brief references to opportunities to expand living/learning opportunities outside of the classroom, augment student internships, collaborate with student organizations, and "build a culture of sustainability at UCSC" (Sustainability Office mission statement 2011).

Key elements had been absent up to this point within UC's policy and UCSC's Campus Sustainability Plans—most notably, there were no references to diversity, equity, inclusion, or multi-culturally relevant approaches to caring for the environment. "Building a culture of sustainability" at that time would have been more accurately phrased as "building a culture of sustainable operations and learning."

As in many prior situations for the UC, it was ultimately the students who would be the catalysts for change. The Student Environmental Center's annual Blueprint for a Sustainable Campus was expanding to include visions and actions for social and environmental justice. Furthermore, as our student population became increasingly diverse, it was difficult to ignore the glaring fact that sustainability spaces across campus academics and organizations were occupied by predominantly white professional staff and faculty members. Apart from occasional references to the "social" leg of the traditional "3-legged stool" model of sustainability, issues related to diversity, equity and inclusion were not at the forefront of discussion for many staff and faculty members working on sustainability.

In my personal experience during that time period, many of my sustainability colleagues assumed that any adverse environmental impacts and disparities were on the basis of socioeconomic status rather than with race, ethnicity and culture. Deeper issues of environmental injustice and the historic racism of the environmental conservation movement were often further removed from their understanding, and sometimes elicited a defensive response when brought up by colleagues and/or students. In retrospect, I can also recognize my own hesitation to acknowledge the depth of these issues. I now see how our lack of understanding as a predominantly white field of environmental professionals has had adverse effects on our efforts to successfully engage with our increasingly diverse student body. As a cisgendered white woman, I had to acknowledge that there was a level of personal growth and development that I had to be willing to take on for myself, *and* encourage amongst my colleagues, in order to successfully broaden the lens of sustainability goals and planning beyond its traditional operational focus. I have also had to embrace that learning, growth and relationship-building within diversity, equity and inclusion work is lifelong.

By 2015, I was growing aware that the sustainability field had a problem. Every single environmental space I participated in with the local community and at the university was vast-majority white, if not *all* white. But like many colleagues who were coming to the same realization, I did not know how to start doing something about it without inadvertently tokenizing folks by bringing them into spaces that were not ready to effectively embrace different cultural perspectives. The students and staff who collaborated early on to form what would eventually become

PoCSC would be instrumental in challenging UCSC's predominantly white campus culture of sustainability, and their student-led events, workshops and social media campaigns started to show me what new and innovative approaches could look like. I stepped out of my comfort zone and started to "show up" in these spaces, even though I did not have much to say yet. I showed up, I was open, I listened, I thanked people for sharing their experiences, and offered myself as a resource if I could be of help. Small successes started to build upon each other in the form of learning, asking critical questions, centering student experiences and building relationships across campus.

As the new Sustainability Director, I had to be willing to turn the lens inward and start asking critical questions about my own organization to determine what we needed to do better to become more inclusive. I had to be willing to re-envision the next Campus Sustainability Plan to center inclusive sustainability in addition to operations, which was successfully incorporated by 2017 (University of California, Santa Cruz, 2017). I had to evaluate our staff and student hiring practices for bias. I also had to be willing to invite campus administrators who had already espoused "values of social and environmental justice" to make space to listen to student, faculty and staff concerns about microaggressions that were being experienced on campus in predominantly white environmental spaces. And when campus administrators proposed traditional organizational models such as a top-down "task force" to address diversity and sustainability, I was appreciative that they were open to feedback from students about how the institutional structures themselves can be problematic.

The Executive Vice Chancellor and Vice Chancellor for Business and Administrative Services were willing to pivot and offered one-time resources to support the emerging model of PoCSC that re-centered student voices and experiences and put leadership within Black, Indigenous, and people of color (BIPOC) centered campus spaces. In my anecdotal experience with sustainability colleagues at other higher education institutions, it can be difficult to obtain this level of acknowledgment and support from high-level campus administrators for this type of work, which is another testament to UCSC's long-standing commitment to both sustainability and social justice. Though we had to be creative to piece together funding in subsequent years, the administration's early support for PoCSC helped UCSC stand out as a leader in the development and implementation of inclusive sustainability.

Rebecca Hernandez

My work as a scholar/practitioner has led me down a winding road, one that has allowed for me to learn in a variety of ways, in different settings and with a wide spectrum of educators. The research and writing I have done in the past examines notions of personal and group identity vis-à-vis the arts. And I have specialized in the scholarship of the convoluted and biased practices of collecting, archiving, and exhibiting Native history and art in museums, libraries and university research and archives. After completing my PhD, I served as Assistant Director of the American Indian Studies Center at UCLA where a large part of my work centered on engaging with faculty and graduate students whose research focused on American Indians.

I arrived at UCSC in 2014 and served as director of the American Indian Resource Center (AIRC) through 2021. The Resource Centers at UCSC[1] are counter-spaces for students who are queer, womxn and BIPOC. They work independently and collectively to positively impact institutional policies and campus climate. The Resource Centers engage the campus on issues and challenges facing our communities and provide programs, physical spaces and services to foster students' academic, personal, professional

[1]There are six Resource Centers at UC Santa Cruz: American Indian Resource Center (AIRC); African American Resource and Cultural Center (AARCC); Asian American / Pacific Islander Resource Center (AA/PIRC); Chicanx Latinx Resource Center (El Centro); Lionel Cantú Gay Lesbian Bisexual Transgender Queer Intersex Resource Center; and Womxn's Center.

growth and leadership development. Students on the campus engage with the Resource Centers in various ways and often build strong relationships with the professional staff, who play an important role in fostering trust and building community on campus and are often called upon to assist and advocate for students in need. Some examples of this are when students speak with Resource Centers' staff about their experiences in the classroom, in dormitories and at off-campus gatherings. Students often share negative comments made by faculty who did not have a clear understanding of the student's community, ethnicity or race. The students would tell us how angry they were about this; how powerless they felt to deal with it directly; and whether this dynamic with the professor might jeopardize their grade, especially for courses required for their major. Students also made clear that living on campus was made difficult by similar experiences with roommates who made rude comments about the food they ate, the age of the technology they used, or the area where they grew up. Students find comfort and refuge in the resource centers; they congregate there with staff members who keep them company and encourage them and students of similar backgrounds.

My role as Director of the AIRC included raising awareness about American Indian history and lifeways both on and off campus by hosting events, speaker series, and presentations. A big part of my job was to ensure that Native students felt a sense of belonging, were seen for who they are today and had a place to be in community with other Native students. Needless to say, the job was different every day, and in my role, I was privy to information shared by students whose experiences were both exciting and inspiring, as well as painful and disappointing.

One day our Managing Director, Nancy Kim, spoke to us about an incident that occurred when a student of color was at our main campus library. After finishing a beverage in the library cafe, student #1 went to dispose of their trash in the bins provided. There were four bins to choose from and they accidentally chose the wrong bin for one item. It happened that student #2 was nearby, witnessed the error and confronted the student about it. However, the approach taken by student #2, who is White, was harsh, rude and demeaning. They humiliated student #1 in a public area, made derogatory and patronizing comments, and asked student #1 why they didn't pay more attention. They said, "It must be because you don't care about the environment." This comment, and the judgment it casts, affects students of color in disparate ways given the heterogeneous racialized experiences vis-a-vis environmentalism. If a Native student had made that mistake, it would have been in direct violation of essentialized notions of the ecologically noble savage (Redford, 2001; Hames, 2007); in contrast, African, Black and Caribbean (ABC) and Asian American and Pacific Islander (AAPI) students are presumed to be indifferent or unconcerned about environmental concerns; and Latinx students are stereotypically seen as perpetrators and/or victims of environmental harm.

Similarly, students from a wide cross-section of identities who were enrolled in Environmental Studies courses and/or engaged with clubs whose focus is sustainability or environmental issues would share that they usually felt uncomfortable and left out of the conversations that took place in the classroom, at club meetings, and events. They were invisibilized due to being silent, or were treated with hostility when they did speak up. This was especially true of students of color who wanted to share ways that they and their families lived sustainably, beyond the typical markers of driving electric cars or eating organic food. One student proudly shared in class that they painted over logos on reusable bags, decorated them with botanical designs and gifted them to friends. Another student in the class commented "that makes no sense," and the other students laughed. Rather than point out how this was a very creative and clever practice, the professor did not acknowledge this exchange between students at all and moved on with the lecture. When students are confronted with dismissive behavior like this, in which efforts to be ecologically conscious are cast as merely "practices of the poor," they disengage. That means everyone in the class loses the opportunity to learn about the wide variety of ways people can contribute, even in small ways, to steward the environment.

As AIRC Director, I was told about many frustrating experiences by students. Far from home, usually the first in their families to attend college, American Indian students (who typically make up less than 1% of the student body at *any* US college or university) in these situations were also weighed down by stereotypical ideas of Native folks, all the usual simplistic imaginings of American Indians being "one with nature." When students, faculty and/or staff learned of a student's Native identity, they were immediately called upon to share about their unique relationship with nature, provide insights into communing with nature in the "Native" way, and give examples of how Indians behave in natural settings when they are alone or with other tribal members. Do Indians "see" things in the natural world that non-Natives can't see? Can they really communicate with animals like in the movies? Are they more at home in nature than anywhere else? Such essentialized and facile assumptions and attributions about American Indians are ubiquitous in our society, but in places of higher learning, one would hope to see greater awareness. Unfortunately, that was rarely the case.

Native college students I worked with often chose not to speak about their ethnicity; if they did, they would have to carry the burden of being the spokesperson for every Native person, living or dead. Adding to the complexity of ignorance in classrooms were the ongoing questions that seek to assess 'authenticity', as though non-Natives are entitled to ask such questions and are qualified to be cultural arbiters. Native students were asked about whether or not they spoke their Indigenous language, grew up on a reservation, or participated in sacred ceremonies. Other times they were asked "how much Indian" they were, and/or told that they didn't "look" Native, causing distress and frustration for even the most confident and well-spoken American Indian students. These hurtful questions and comments by non-Natives about what Natives should know, be, act and look like are big obstacles to success for American Indian college students. Story after story was shared with me about these difficult encounters, especially those experienced in the classroom. I believe this is rooted in the overall ignorance about American Indians in this country. This was particularly true for students who were majoring in earth science, ocean science, astrophysics, environmental and/or sustainability studies. Again, when people knew there was a Native student in a class or in the major, that student was expected to share personal information about themselves, their families and tribes, expectations that are not imposed on non-Native students. The fantasy that Native folks understand the natural world in some sort of magical way is at the root of this expectation, and constitutes yet another realm in which they are prompted to demonstrate their 'authenticity' for the non-Native gaze. These pressures and burdens cause Native students to feel out of place, making it much more difficult to retain them.

Returning to the trash bin story at the start of this section, at that point all the Centers directors decided it was time to work collectively around shifting the discourse about how students of color engage in sustainability efforts. We approached our Race, Class & Ethnicity Specialist and asked if he would take on the task of finding a student who was interested in moving the conversation forward. Soon after, Adriana Renteria was hired and the hashtag #POCSustainability was created. Without knowing it at the time, the need for this hashtag and space for sharing was desperately needed; within weeks, the hashtag began to take on a life of its own. Students of color on campus eagerly shared the ways that they practiced sustainability, were excited to have conversations about sustainability and to raise awareness about unseen and under-appreciated ways that they engaged in sustainable practices. Reusing plastic food containers such as margarine tubs for leftovers, old tires for raised garden beds, cans for seed starters and washing plastic sandwich bags was just the start.

When it was time to find a home for PoCSC, it seemed a natural fit for the American Indian Resource Center. The reason for this was simple: as Native folks we needed to dispel some of the stereotypes brought up earlier. We wanted to empower American Indian students to speak with confidence about their own, contemporary lived experiences and encourage them to research and present about pressing ecological issues in their communities, whether on the reservation or in urban areas where they lived before coming to college. Additionally, the work of the Resource Centers focuses on the retention of students, and this initiative was one that brought together many students from different backgrounds around a common theme. They found a place where they could share their interests, and help raise awareness about the environmental injustices they personally witnessed in the areas where they grew up and in relationship to BIPOC more generally.

Since PoCSC began, we have made it a point to evolve and try different approaches to ensure that students lead our efforts. That includes hiring between three and five undergraduate students each academic year to serve as interns focused exclusively on the work of PoCSC. In the past we have also had a Program Coordinator, with Adriana as our first. More recently, a graduate student from Environmental Studies served as the coordinator for PoCSC, allowing for the group to be led by a fellow student who has expertise in the field. At its core, PoCSC exists so that students—students of color in particular—can (re) shape and influence the narrative around sustainability. Far from not caring or not understanding environmental issues, these students are mindful and concerned. They, more than any generation before, recognize the urgency of the situation, know their lives and those of their children are now altered in ways unimaginable 50 years ago. Providing a space for them to learn and share outside the classroom allows for students to think about solutions for the common good.

Adriana Renteria

In 2015, the People of Color Sustainability Collective emerged after many years of student discontent with the mainstream campus environmental culture. Certain facets of this discontent are visible through the notes gathered as part of discussions for the annual Blueprint for a Sustainable Campus (Lippus et al., 2013). The Blueprint is a student-authored document published by the Student Environmental Center (SEC) that acts as a counterpart to the Sustainability Office's Campus Sustainability Plan. The SEC has issued a new version of the Blueprint every year since 2002. The goal of this document is to put forth short- and long-term goals for sustainability on campus and beyond. Every year, the SEC hosts small group discussions to develop visions and actions focused on the different topics of the Blueprint. The Blueprint influences courses offered through the Education for a Sustainable Living Program (ESLP), an interdisciplinary, student-facilitated experiential learning program that fosters wider discussions of sustainability within academia and greater society. The Blueprint also distributes funding through the Campus Sustainability Council (CSC) to support sustainability projects pitched by other registered, campus-based student organizations that align with the visions and actions described in the Blueprint. These various efforts additionally facilitate greater collaboration between students, the administration, staff, faculty, and the community. Most importantly, the Blueprint offers an honest and critical reflection of campus sustainability from a student-led perspective. It is through the notes and visions described in the Blueprint that we can see the evolution of campus culture and student priorities.

Recent Blueprint notes highlight a general lack of awareness about social and environmental justice topics and lack of understanding that environmental issues are fundamentally inextricable and linked to social justice. The first Blueprint discussion that addressed social justice took place in 2009 where five participants shared reflections including: "Current Situation: Attendees know little about official policies and the history of social justice at UCSC;" "Social Justice activism and consciousness must be emphasized as something positive– making activism not a dirty word"; and "Is equity the same as equality?" (Carrillo-Mandel, 2009). The discussions in the 2010 Blueprint notes included mention of both social and environmental justice, but solutions focused mainly on building connections with existing social justice movements/organizations—describing them as separate and distinct from environmental movements and organizations. The 2011 Blueprint discussion notes begin to acknowledge and discuss the changing racial and ethnic makeup of the university to increasingly reflect state demographics, but focus on specific spaces, such as African American themed housing on campus, as well as the funding opportunities linked to becoming a "Latino supporting institution" (i.e., Hispanic Serving Institution) (Pitts, 2011). The 2012 Blueprint notes (Milio et al., 2012) are the first that explicitly acknowledge the divide between ethnic/identity-based groups and environmental groups but describe them as two separate efforts that should be collaborating:

> Why bridge environmental and social justice movements? All these groups are seeking one common goal of creating a better campus. There are two sides of the coin; that is, sustainability depends upon social equality. Because in order to achieve sustainability, everyone needs to 'pitch in'. Furthermore, the environmental movement has, in history, oppressed indigenous, low-income people, that racism was justified in the process. As such, environmentalism and social justice ought to cooperate to promote liberation. They are inter-related because both inter-lap in injustices committed (p. 36).

In 2013, I joined the SEC and was the co-coordinator for the Earth Summit and the Blueprint. I organized and co-led the 2013 Blueprint discussions and most of the notes on social and environmental justice were similar to those of the 2012 Blueprint notes: social and environmental movements were described as separate and distinct, yet needing to be bridged. For example, issues included the need to implement ESLP classes focusing on Environmental/Social Justice; to incorporate "multicultural programming" and "cultural days"; and to invite low-income, BIPOC on campus—namely the "bus driver, janitor, food-service, etc. employee"—to share their stories and struggles at "Alliance/Relationship-building events".

This perspective that the environmental movement should be "diversified" assumes that efforts to protect and celebrate the environment are led by, and for, white members who are looking to "add" members of color to their existing efforts. This results in a lack of recognition of existing members of color, tokenizes BIPOC, dismisses their environmental contributions, and reinforces an "Us vs Them" power dynamic. This creates a harmful dynamic and culture that is not comfortable or safe for many BIPOC.

As an active student leader involved in both environmental and identity-based campus groups, I shared this frustration and the desire to see a transformation in our campus culture towards inclusivity and justice. During my time in the student environmental movement on-campus and as an Environmental Studies major, there were many times that I felt like an outsider who did not belong. I remember a time another student made a comment about my stained plastic tupperware, saying that I should get a glass container because glass was better for the environment than plastic. I remember reflecting that my tupperware were actually passed down from my mom, who got them from my Tía, who got them at a yardsale, and in that way, keeping my stained plastic tupperware may have been more environmentally conscious than purchasing a new glass set. I also remember my peers often having conversations around camping and hiking in which I could not engage. I had not gone camping or hiking until I arrived in college because for my family, many of whom work in agriculture, spending time camping was not a common form of leisure. I also remember a fellow classmate asking our Environmental Studies professor why we did not have authors of color included in our class syllabus to which our professor responded he didn't know of authors of color who wrote on the subject (water policy). And as we were walking out of class, I remember overhearing a student say to another student "things don't always have to be about race." I remember how these comments and conversations made me feel insecure and like an outsider. I was also part of Hermanas Unidas, a Chicana/Latina retention student organization that *did* allow me to experience a feeling of safety and the invitation to show up as my authentic self. The feeling of community I felt when I was around other members of Hermanas Unidas was the feeling I craved in the environmental spaces of which I was also part.

In 2014, as we were planning the Earth Summit and associated Blueprint discussions, a group of other student leaders and I decided to host a students of color environmental caucus at the annual Earth Summit. This was the first time that students of color had created a closed space to discuss amongst themselves. After the caucus, we proudly shared our discussion notes with the larger Earth Summit attendees. It was during this caucus discussion that I finally felt a taste of the sense of community I had been seeking. It was also during this caucus that I first met Karim Ahmath, the Race, Class, and Ethnicity Specialist for the Ethnic Resource Centers, who had attended the caucus to listen to students. I had a follow up meeting with Karim after the Earth Summit to explain the challenges we were experiencing and our vision for change. A few weeks later, Karim offered me an opportunity to work as an intern with the Ethnic Resource Centers to carry this work forward. From this place, I had the privilege of playing a key role in the formation of the People of Color Sustainability Collective, first as an student intern and later as the Collective's first Program Coordinator.

The People of Color Sustainability Collective

As described in the Introduction of this edited volume, the People of Color Sustainability Collective, or PoCSC, emerged from a growing desire by students of color to create a welcoming space for themselves outside of their tokenization within mainstream environmentalism. Rather than viewing justice and culture as separate from the environment, the initiative recognizes that in many ways, our cultures and our identities deeply influence our relationships with the natural environment. The People of Color Sustainability Collective coined the term ***inclusive sustainability*** as an effort to create community, support student retention, and expand our collective understanding of sustainability. By creating student spaces that embrace inclusive sustainability, PoCSC has been building community and deepening our understanding of sustainability. Below are elements that encompass the core tenants of inclusive sustainability and the benefits of this approach to relating to the environment.

Embracing Cultural Approaches to Caring for the Environment The People of Color Sustainability Collective first emerged as the hashtag: #POCSustainability. The purpose of the hashtag and social media campaign was to start a dialogue about the many different ways that we, our communities and cultures, contribute to sustainability, which often go under-recognized by mainstream environmentalism. The hashtag created a digital platform for students to share their experiences and (re)claim the often sustainable practices of their families and cultures. At several tabling events, the hashtag was used to ask event attendees questions like: how do you or your family practice sustainability, and what are some cultural practices you or your family do to care for the environment? Event attendees would share their verbal or written responses online using the hashtag. What emerged from these conversations was an understanding that many cultural practices include elements of conservation, recycling, reducing waste, or honoring the environment. Some examples students shared were: reusing empty butter containers to use as tupperware; wearing hand-me-down or thrifted clothes; reusing old, stale tortillas to make chilaquiles; and many more.

Prior to this hashtag, there did not exist many spaces at UC Santa Cruz for students to speak about these practices with others who may have had similar experiences or cultural practices. As student responses grew, PoCSC crystallized and began incorporating more programming beyond the hashtag #POCSustainability. Events were organized to continue this conversation to give students a space to talk about their experiences in person. We held events like "Re-Imagining the Environmentalist" where we students discussed stereotypes related to identities and the environmental movement; or "Mapping our Journey" where students painted images of landscapes, food, and cultural practices that were significant in their life trajectory and then shared their stories and paintings with each other. We also developed and led a course for faculty and staff through our Office of Diversity, Equity, and Inclusion's Diversity and Inclusion Certificate Program called "Intersections Between Diversity and the Environment."

Validating All Our Identities Inclusivity is a critical element of the community PoCSC works to create and was one of the most important elements for me personally. As a student, I (Adriana) often felt as though I needed to code switch between different groups I was a part of. When I was with student environmental groups, I could connect with others about our passion for nature, protecting the environment, or about the courses we were taking. When I was with my Chicanx/Latinx student groups, I could connect with others who shared a similar upbringing and perspective of the world and who may also be feeling homesick or out of place on campus and with other peer groups. I yearned for a group where I could bring all aspects of my identity together without needing to code switch. Having to internally negotiate which parts of myself I felt safe sharing with others was exhausting and isolating. Helping establish and grow PoCSC allowed me the opportunity to create the community I had been searching for and craving. PoCSC filled this gap by explicitly acknowledging that our cultures, upbringings, and intersectional identities deeply influence our understanding of, and relationship with, the environment. It created spaces to explore our connection to the environment, removing the embarrassment of not having grown up doing traditional environmental activities like camping, rafting, hiking, etc. For those of us who grew up in urban or highly densely populated areas, these spaces allowed us to revisit and recognize the nature that was present, evoking memories of familiar trees from our apartment complexes or the potted plants our grandmothers kept. Inclusive group spaces free from shame, judgment, embarrassment, or limiting expectations, invited us to bring our whole selves and all our identities to the conversation, to show up authentically and freely express ourselves. These generative discussions enabled us to redefine sustainability and our relationship to the environment, both for ourselves and for the collective.

Disrupting the University as a Gatekeeper of Knowledge Another element of inclusive sustainability is the practice of disrupting the idea that the university is the sole keeper of knowledge and the de facto expert on sustainability or the environment. The traditional view of universities assumes that we enter the university as empty vessels, ready to be filled up with the ideas, perspectives, and world views of professors who are the experts. When professors and their curriculums do not reflect the diversity of the students they teach, students are left without examples that reflect their experiences.

For many first-generation college students of color, entering the field of environmental studies left little room for the lessons and wisdom our parents and elders taught us and for our lived experiences. Furthermore, concepts like Indigenous land management or agricultural practices were often appropriated and rebranded as having been developed by the white men teaching them, without acknowledging the historical and cultural context of those practices, or if acknowledged, doing so in ways that simplified, essentialized and romanticized Indigenous peoples. PoCSC subverted the gatekeeping of knowledge by inviting students to share their lived experiences or the stories passed down by their elders, and by bringing recognition to the historical contributions BIPOC have made towards sustainability. PoCSC staff and student interns further disrupted this idea by undertaking research on the campus environmental climate. Chapters in Part II of this volume share some of the results of this research.

Promoting Strategies for Student Retention, Success, and Community Building The reason for housing PoCSC within the Ethnic Resource Centers, and in collaboration with College Nine, John R. Lewis College, and the UCSC

Sustainability Office, stems from the role the Collective plays in supporting student retention, success, and community building. The Resource Center Directors had heard many stories from students of color majoring in Environmental Studies and quickly recognized this was an issue affecting student retention and well-being. They pushed to dedicate resources to address this issue and support the creation of an inclusive community, something these units were uniquely designed to do. Like Dr. Hernandez describes earlier in the chapter, this was a specifically necessary tool for the retention of Native students who often experienced microaggressions in environmental spaces.

Advancing Environmental Justice Statewide The work of PoCSC, and of similar models of inclusive sustainability programming, have far reaching implications even beyond the confines of the university campus. From my current vantage point as the Director of the Office of Public Participation at the State Water Resources Control Board, I (Adriana) can better appreciate and recognize the critical role that student retention organizations have on helping create a diverse environmental workforce.

The State Water Resources Control Board and the nine Regional Water Quality Control Boards, collectively known as the Water Boards, are state agencies responsible for preserving, enhancing, and restoring the quality of California's water resources and drinking water for the protection of the environment, public health, and all beneficial uses. The Office of Public Participation was created to ensure the Water Boards' decision-making processes involve the public and uplift perspectives from environmental justice communities and California Native American Tribes. Our environmental justice work includes providing language access, facilitating community engagement to advance safe drinking water solutions, facilitating public engagement in water decision making processes, and coordinating government-to-government engagement between the Water Boards and California Native American Tribes. The last couple of years I have also played a leadership role in advancing racial equity at the Water Boards through membership in the Water Boards' Racial Equity Team where we are working towards a future where: (1) race no longer predicts a person's access to water or the quality of water resources they receive; (2) race does not predict professional outcomes for our employees; and (3) we consistently consider racial equity impacts before we make decisions.

Our work to advance racial equity and environmental justice is data driven. The California Environmental Protection Agency's 2021 Pollution and Prejudice Story map demonstrates that historically redlined neighborhoods are "generally associated with worse environmental conditions and greater population vulnerability to the effects of pollution today." BIPOC are over-represented in the neighborhoods that are the most environmentally degraded and are still experiencing severe racial wealth gaps caused by redlining and other land-use practices (Altare et al., 2021). The Office of Environmental Health Hazard Assessment's (OEHHA's) CalEnviroScreen mapping tool identifies communities that are disproportionately impacted by a combination of environmental stressors and socioeconomic disadvantages. The tool's 2021 update reveals that the top 10% of least polluted neighborhoods are 72% white, while the top 10% of most polluted neighborhoods are 89% BIPOC (Office of Environmental Health Hazard Assessment, 2021).

Even though the data show that the most polluted neighborhoods are BIPOC, the state's environmental workforce does not reflect the same diversity of the communities it serves. This can be seen in the workforce demographics of the Water Boards. The state of California workforce census data from 2020 show that 43% of the state's population is white, yet about 56% of the Water Boards' workforce and 68% of the Water Boards' management is white. Only 42% of the Water Boards' workforce and 32% of its management are BIPOC compared to 63% of the state's population (California Environmental Protection Agency Racial Equity Team, 2020).

Inadequate workforce diversity is a problem because it leads to less innovation and potentially less meaningful connections to communities of color (Hofstra et al., 2020). To address the state's environmental injustices, it is important for the environmental workforce to look like the communities it serves, to have personally experienced environmental injustices, to understand and speak the languages of the community, and to be trusted by the community. Expanding the environmental workforce to meet this goal begins with an inclusive K-12 and university education and culture. Expanding our understanding of sustainability and of the environment creates a welcoming opportunity for students to explore whether a career protecting, restoring, or celebrating the environment is something they would like to pursue. By creating this opportunity and redefining what it looks like to work in an environmental field, we are able to support the expansion of an environmental workforce that reflects the diversity of the state of California and that shares in the lived experiences of communities currently experiencing environmental injustices. PoCSC is helping prepare environmental leaders who are BIPOC in order to meet this challenge and to advance environmental justice. It is through creating student retention spaces that center cultural understandings of the environment and inclusive community building that we will be able to rise to the challenge of addressing climate change and protect the planet for the benefit of present and future generations to come.

Conclusion

Since 2015, the People of Color Sustainability Collective has sustained its groundbreaking collaboration and succeeded in creating inclusive sustainability spaces for students of color to thrive. We have learned a great deal and continue to grow in our understanding of how to best serve and support student success through inclusive sustainability.

If you are interested in implementing a similar model on your university campus, here are a few lessons learned from our journey:

- Listen and follow the leadership of students of color.
- Personal awareness of race and privilege and a commitment to transformational growth is required of all people assuming leadership roles.
- Expanding collaborations beyond traditional partners requires intentional and on-going relationship building.
- Find high-level administrators who appreciate your efforts and communicate with them regularly to share successes.
- Host events where students can present on various environmental and sustainability topics and be intentional about promoting those events widely.
- Utilize art and storytelling as mediums for self-expression and community building.
- Critically evaluate the inclusivity of your institution's or department's stated and enacted values, programs, and hiring processes.
- Adopt the mindset of being on a continual learning journey. One approach to create that mindset is through the exploration of critical questions, such as the ones offered in the Self-Assessment for Diversity, Inclusion and Sustainability developed by PoCSC co-founders (see Appendix A).

We encourage readers to reflect on the opportunities available to create, or support, similar efforts on their own university campuses. The self-assessment available in the Appendix can be used to evaluate yourself and your institution. The assessment is intended to help identify areas of growth to focus on when embarking on your journey to support an inclusive and welcoming environmental culture on your university campuses. This work requires consistency and commitment. It can be challenging and frustrating; however, it's also incredibly rewarding and inspiring.

References

Altare, D. et al. (2021, April 19). CalEPA Story Map. Pollution and Prejudice: Redlining and Environmental Injustice in California. https://storymaps.arcgis.com/stories/f167b251809c43778a2f9f040f43d2f5

California Environmental Protection Agency. (2021). Pollution and prejudice. Retrieved October 2, 2022 from https://storymaps.arcgis.com/stories/f167b251809c43778a2f9f040f43d2f5

California Environmental Protection Agency Racial Equity Team. (2020). Racial Equity Data Resources. Retrieved October 2, 2022 from https://caleparacialequity.github.io/racial-equity-data-resources

Carrillo-Mandel, S. (2009). 2009 blueprint for a sustainable campus. Student Environmental Center, UC Santa Cruz. Retrieved October 2, 2022 from https://static1.squarespace.com/static/56c3bc72b09f95c8b3b19126/t/5b836c5603ce64a5a9b66df9/1535339619301/blueprint_for_a_sustainable_campus_2009+%281%29.pdf

Dynes, R. (2007, August 28). University of California Office of the President (UCOP) Leadership in Engineering and Environmental Design for Existing Buildings – Operations and Maintenance (LEED-EBOM) ceremony. [Speech].

Hames, R. (2007). The ecologically noble savage debate. *Annual Review of Anthropology, 36*, 117–190.

Hofstra, B., Kulkarni, V., Munoz-Najar Galvez, S., He, B., Jurafsky, D., & McFarland, D. A. (2020). *The diversity–innovation paradox in science.* Columbia University. https://doi.org/10.1073/pnas.1915378117

Lippus, K., Renteria, A., Yaros, C., & Mukerji, R. (2013). 2013 blueprint for a sustainable campus. Student Environmental Center, UC Santa Cruz. Retrieved October 2, 2022 from https://static1.squarespace.com/static/56c3bc72b09f95c8b3b19126/t/5b836a0b562fa70992ef9863/1535339061116/blueprint_2013-14_final+%282%29.pdf

Milio, E., Bobro, C., McKee, C., & Nelson, C. (2012). 2012 blueprint for a sustainable campus. Student Environmental Center, UC Santa Cruz. Retrieved October 2, 2022 from https://static1.squarespace.com/static/56c3bc72b09f95c8b3b19126/t/5b8368d703ce64a5a9b64e1b/1535338722730/officialblueprint2012+%281%29.pdf

Office of Environmental Health Hazard and Assessment. (2021). CalEnviroScreen 3.0. Retrieved October 2, 2022 from https://oehha.ca.gov/calenviroscreen

Pitts, T. (2011). 2011 blueprint for a sustainable campus. Student Environmental Center, UC Santa Cruz. Retrieved October 2, 2022 from https://static1.squarespace.com/static/56c3bc72b09f95c8b3b19126/t/59f8088a53450a4c5a506b04/1509427367289/2011+sec_blueprint_for_a_sustainable_campus_2011.pdf

Redford, K. (2001). The ecologically noble savage. *Orion, 9*, 24–29.

Skallerup, L. B. & Fisher, J. P. (2021, June 23). Student success this fall will depend on faculty-staff cooperation. The Chroniclve of Higher Education. https://www.chronicle.com/article/student-success-this-fall-will-depend-on-faculty-staff-cooperation

UCSC Public Affairs. (2013). UCSC's founding principles featured in 'Time,' 'New York Times Magazine' ads. Retrieved October 13, 2022 from https://news.ucsc.edu/2013/10/questioning-authority.html

University of California. (2021). UC Sustainable Practices Policy. Retrieved October 1, 2021 from https://www.ucop.edu/sustainability/policy-areas/index.html

University of California, Santa Cruz. (2014). Chancellor's Vision. Retrieved May 5, 2018 from https://chancellor.ucsc.edu/

University of California, Santa Cruz. (2017). 2017–2022 Campus Sustainability Plan. Retrieved October 14, 2021 from https://sustainabilityplan.ucsc.edu/

University of California, Santa Cruz. (2022). Sustainability Office Mission Statement. Retrieved October 1, 2021 from https://sustainability.ucsc.edu/about/index.html

Elida Erickson joined the University of California, Santa Cruz campus in 2005, and the Sustainability Office in 2011. Over the years, she has collaborated with the local Santa Cruz community, students, faculty and staff to support the goal of Zero Waste by 2020, as well as reduce campus water usage by 25% at the height of the California statewide drought in 2014–2015. In her current role as Sustainability Director, she is a strong advocate for student engagement and professional growth, and is passionate about challenging the sustainability movement to open up to multi-culturally relevant interpretations of how to care for the environment. She played a foundational role in the development of UCSC's award-winning People of Color Sustainability Collective (PoCSC), which coined and defined the concept of "inclusive sustainability" in 2015. Elida holds a Bachelor of Arts degree from Loyola University New Orleans, and Master of Science in Higher Education and Student Affairs Administration from Indiana University.

Rebecca S. Hernandez is Mescalero/Warm Springs Apache and Mexican American. She is the inaugural Community Archivist at the University of California, Santa Cruz University Library where she partners with local stakeholders to promote the acquisition, preservation, and use of archival materials that document the communities of Santa Cruz County. Rebecca served as Director of the UCSC American Indian Resource Center from 2014–2021 where she worked with the team to create collaborative programs across campus including the People of Color Sustainability Collective and the UCSC Missing and Murdered Indigenous Women and Girls Series. Rebecca holds an MFA in Design from CSU Fullerton, is a graduate of the UCLA American Indian Studies MA Program and earned her PhD in American Studies at the University of New Mexico.

Adriana Renteria (she/they) is the Director of the Office of Public Participation and Tribal Liaison at the California State Water Resources Control Board. Adriana leads a team who work to ensure the Water Boards' decision-making processes involve the public and uplift perspectives from environmental justice communities and California Native American Tribes. Previously, Adriana was Regional Water Manager at the Community Water Center where they worked on local efforts to protect safe drinking water and domestic well users through the implementation of the Sustainable Groundwater Management Act, and where they also worked to endorse and support leaders running in San Joaquin Valley local water districts elections. Adriana was a founding member and the first Program Coordinator for the People of Color Sustainability Collective at the University of California, Santa Cruz's American Indian Resource Center. They have years of experience supporting community engagement in water governance and they have developed and facilitated workshops, programs, and events related to environmental justice, food justice, Human Right to Water, groundwater management, racial equity, tribal affairs, local water district elections, language access, and more. Adriana graduated from UCSC with a B.A. in Environmental Studies and Economics.

Student Voices on Environmental Spaces and Experiences in Higher Education

2

Kimberly Dare and Riri Shibata

Abstract

This chapter highlights the voices of four Black, Indigenous, and People of Color (BIPOC) affiliates of the Environmental Studies (ENVS) department to discuss topics of marginalization and dissociation of the academic environment at the University of California, Santa Cruz (UCSC). While UCSC is prominently known for its liberal ideals, sustainability initiatives, and social justice, these individuals argue that the ENVS department is far from the picturesque vision that people paint it to be. Specifically, ENVS is a department where whiteness is overrepresented demographically and curricularly; as such, these roundtable discussants share that the student experience often includes isolation and disappointment. We argue that there needs to be more attention on empowering BIPOC students through a critical lens in hopes of attracting and retaining more BIPOC individuals in the environmental field.

K. Dare (✉)
University of California, Santa Cruz, CA, USA
e-mail: kdare@ucsc.edu

R. Shibata
California State University, San Francisco, CA, USA
e-mail: rshibata@sfsu.edu

Keywords

Environmental studies · BIPOC · Sense of belonging · Student experience · Student development theory

Introduction

The transition between high school to college can come with many challenges. Moving away from home to navigate a new campus can create an onslaught of change during a pivotal period of their adolescence, resulting in many students struggling to graduate or even make it through their first year. According to the National Student Clearinghouse Research Center (NSCRC) report (2022), as of July 2020, there are approximately 39 million Americans who identify as part of the "some college, no credential" population (Causey et al., 2022: 4). Another report in 2022 by the NSCRC estimated that more than 25% of first-time freshmen do not return for their second year (Gardner, 2022: 4). These rates are particularly high among Black, Latinx, and first-generation students (Kirp, 2019). From financial to social concerns, a variety of factors contribute to these numbers.

According to the 2019 report released by the American Council on Education (ACE) on "Race and Ethnicity in Higher Education," about 45 percent of undergraduate students in 2015–16

F. Lu, E. Murai (eds.), *Critical Campus Sustainabilities*, Sustainable Development Goals Series,
https://doi.org/10.1007/978-3-031-30929-8_2

identified as a person of color or ethnicity other than white, which has grown significantly from the 1995–96 study that reported about 29.6 percent students of color (American Council on Education, 2019). In addition, the National Association of Student Personnel Administrators (NASPA) reported in their 2019 First-generation Student Success data that a third of today's college students are first in their family to step foot on a college campus as a student (Whitley et al., 2019: 6). From maintaining jobs and balancing family obligations, first-generation college students undergo a multitude of challenges while trying to obtain their degree. As a result, only 27% will attain their degrees within 4 years—markedly lagging behind their continuing generation peers (Whitley et al., 2019: 6).

The global pandemic has disproportionately impacted these communities as they entered college during the peak of the pandemic (Grim et al., 2022). They also tend to be further marginalized, isolated and pushed out by institutions and systems that do not attend to their unique needs. Research from several reports examining students in grades K-12 through college show growing evidence for the need for faculty diversity across all levels of education systems. ACE's "Race and Ethnicity in Higher Education," explains it as such: "… faculty diversity positively affects Black students in ways that parallel how increases in female professors in STEM subjects influence female students' choice of major (Bristol & Martin-Fernandez, 2019; Carrell et al., 2010; Price, 2010) …[how] teachers can affect student leadership qualities, motivation, socio-emotional development, and career choice should not be overlooked (Johnson, 2020)" (52). As other chapters in this book show, disciplines like Environmental Studies and Sustainability Studies have centered individualistic, lifestyle and techno-managerial definitions in their approaches to the environment, overlooking questions of social justice, race, equity and inclusion, which in turn have led to issues of under-representation and racial gaps in both disciplines. If faculty and administrators wish to attract and retain students from a diverse range of backgrounds into their departments and into the career pipeline, they must address the histories of exclusion and elitism that underlie their disciplinary foundations and center the needs and perspectives of underrepresented groups. Without challenging the dominant racial dynamics that shape these fields and make meaningful efforts to overcome them, both systems of higher education and specific disciplines will continue to perpetuate a field that is predominantly white.

In this chapter we center Black, Indigenous, and People of Color (BIPOC) student voices through a roundtable discussion on environmental themes as a way to showcase the environmental perspectives of students who are often peripheralized and tokenized in these debates. Throughout the chapter, we use the term "BIPOC" to describe the participants for inclusivity purposes.[1] This candid conversation among four BIPOC affiliates from the Environmental Studies department (ENVS) at the University of California, Santa Cruz (UCSC) discusses sensitive topics such as dealing with isolation, witnessing a lack of racial and ethnic representation in the department, and building agency and resilience in the face of marginality. The voices of these BIPOC students offer a valuable lens into understanding the ways theories and practices of student development get shortchanged in disciplines that ignore these issues. Through their discussion, it will be clear that ENVS—a field that is predominantly white in terms of faculty and student representation—needs to make strides to create a welcoming and equitable academic environment for all students, but especially those who are marginalized in terms of race and first-generation status.

Who Are We?

Kimberly (Kimie) Dare

I graduated from UCSC with an Environmental Studies major and Education minor in 2021.

[1]"BIPOC" has been coined to elevate the voices of Black and Indigenous people whose groups have been historically marginalized and excluded (Garcia, 2020).

Similar to the statistics outlined above, my personal transition to UCSC was an immense challenge. Coming from a predominantly Asian city in the Bay Area, I experienced culture shock and isolation as a Chinese American navigating the cultural homogeneity of the ENVS department. I yearned to withdraw my first academic quarter. While my first year posed many challenges, in my sophomore year, I became more acclimated to my environment through active involvement with internships and teaching opportunities. As I immersed myself in my studies and my academic network expanded, I gratefully received support and guidance from all tiers on the university campus—peers, graduate students, staff, and faculty—which alleviated my negative experiences and strengthened my sense of belonging. The people who graciously mentored me, many of whom were similarly BIPOC, exposed me to new issues and viewpoints, and refined my perspectives on race and ethnicity to be more critical. Through a multitude of conversations, I became more versed on racial topics and was inspired to undertake a departmental-wide research project to further explore the educational setting for BIPOC students and find potential areas for reform.

Thus, in my senior year, I wrote an award-winning thesis (Dare, 2021) that was later revised and published in the journal, *Global Environmental Justice* (Dare, 2022). For my thesis, I conducted interviews with BIPOC students and faculty about ENVS' climate and student success. In my research, I found that BIPOC students in ENVS have a less than moderate sense of belonging; value learning about environmental justice and the socio-cultural aspects of environmentalism; and desire seeing their racial/ethnic identities represented within the department. With direct insights from 21 interviewees and a wide swath of academic research, I developed a list of recommendations to foster greater belonging, such as amplifying students' concerns, establishing opportunities for social events, and implementing a more inclusive pedagogy. My research and related policy memo were widely disseminated and garnered attention departmentally and campus-wide. I received the 2021 Deans' and Chancellor's Awards, two of the most prestigious campus academic awards for undergraduates. Furthermore, I presented my findings at the 2022 American Association of Geographers' annual conference where I received the Gail Hobbs Student Paper Competition Award for the Geography Education Specialty Group. Beyond these accomplishments, the interviewees were appreciative that their voices would be heard and their input would impact future cohorts of BIPOC students' experiences, making the project personally rewarding for me. The success of my thesis and enthusiasm from fellow students to practitioners to hear BIPOC students' voices inspired me to continue my research by facilitating this virtual roundtable discussion in the summer of 2021.

Riri Shibata

I am an alumna of UCSC from the Environmental Studies and Legal Studies departments. During my time as an undergraduate student, I completed and published two award-winning theses under the mentorship of faculty, peers, and staff in both departments. As someone who emigrated from Japan at a young age, I am the product of mentors and advisers of color who supported my academic path and personal growth. Their guidance helped me develop the confidence to find my passion and pursue it whole-heartedly, and find a sense of belonging in both of my majors. Both theses I completed were extremely fulfilling to my undergraduate career, as I had the chance to bring attention to the daily, lived experiences of the Okinawan people in my home country in a context of decades-long U.S. military occupation. I began this work as part of the UCSC Challenge Program, in which first-year students take research methods courses and engage with faculty mentors from all disciplines. Through the mentorship of faculty members, financial support from grants and scholarships, and my dedication to this research project, I was able to complete my research study, present my study at four conferences, and receive the 2018 Chancellor's Award and 2018 Deans' Awards. I learned the significance of research, both its ability to shed light on pressing issues, to amplify under-represented voices, and to foster a sense of belonging for students like myself.

As a result of this strong support system I had, primarily from a diverse set of faculty, staff and students of color across campus, I devoted my post-graduate career to supporting first-generation students and students of color to navigate their own journeys through college. After graduation, I served as the ENVS Undergraduate Program Coordinator and Academic Adviser for two and a half years, a period that included the onset of the pandemic and its profound impacts on education. It was during this time, as a relative newcomer to the profession, I realized for the first time that academic staff, who are the backbone in every higher education setting, struggle with ongoing issues like low and stagnant pay, critical understaffing, high turnover, high student to staff ratios, inconsistent health care benefits, and other systemic problems that compromise one's financial stability while adding stress and inefficiency to workplaces.

In ENVS in particular, the department enrolled over 500 undergraduate students in the Fall of 2021, 263 of whom are non-white (Office of Institutional Research, Assessment, and Policy Studies [IRAPS], 2021). According to the 2011 National Academic Advising Association (NACADA) survey, the median number of advisees per advisor for a full-time advisor was 296 students (NACADA, 2011), showing the level at which the ENVS department was understaffed. Although there are no official recommendations given by NACADA for adviser to student ratios "since advising duties vary by institution type and the purpose(s) of the advising office," as the only staff of color and the sole academic adviser in a growing department, I felt an overwhelming pressure to support our growing number of students of color. I wanted to make sure students felt represented and heard, but due to lack of time and resources, I was unable to connect most of them to resources that reflected their interests. All the while, I supported students through campus power outages, Cost of Living Adjustment (COLA) strikes, international student visa policy changes, wildfires that threatened safety and infrastructure, and the rapid transition to remote learning due to COVID-19. Faced with these unprecedented challenges, faculty and staff in ENVS adapted to maintain the continuity of education and ensure our students were supported to the extent possible with stretched resources. During times of uncertainty and crisis, I focused on ensuring that every student who entered our advising space felt heard and left with a clearer path forward. At the outset of the COVID-19 pandemic, I relied upon best practices of academic advising, specifically drawing on student development theories that fostered a sense of belonging to make sure that our department continued to facilitate student success and mental well-being.

As I read through this roundtable discussion and reflect back on my own experience working with the students in the ENVS department, I find myself referring back to the stages of student development and where our department fell behind in creating spaces that could help these students find their passions and achieve their own definitions of success. I also think back on my own experiences as a graduate from ENVS. I discuss my reflections in more detail at the end of the chapter.

Roundtable Methods

Kimie convened and facilitated the roundtable discussions virtually during summer 2021 in two, hour-long sessions. Kimie utilized a focus group format shared by Adamson et al. (2000) in their own roundtable discussion of environmental justice leaders. In order to capture nuanced perspectives, Kimie selected panelists that reflect a diverse range of racial/ethnic backgrounds, gender identities, and level of involvement in ENVS and environmental organizations at UCSC more generally. Recruitment was executed through direct outreach to the targeted individuals. After recruitment, Kimie developed a series of discussion questions to cover a variety of topics on environmentalism as a movement, Environmental Studies as a discipline, and particular issues

BIPOC students encountered at UCSC. Participants weighed in on the final list of topics, reflecting a co-constructive and shared process that can be applied to improve higher education sustainability spaces. This transcript has been edited for length and clarity. Participants also chose pseudonyms for identity protection. The research findings are presented thematically as follows: getting to know each other, mainstream environmentalism, envisioning equity, sustainability on campus, ENVS, and moving forward.

Getting to Know Each Other

Prior to delving into the other five topics, the panelists share their salutations, backgrounds, and campus involvement with each other as they virtually gather for the first time.

Kimie: Could you please introduce yourself?

Maxim: Hi y'all. Maxim, he/him. I graduated in 2020. I am a Latino from LA, born and raised here. My interest in Environmental Studies is urbanism, like urban planning, cities, etc. I worked as an advisor and tutor within the Environmental Studies department.

Marissa: Hello, my name is Marissa. Pronouns—she/her/hers. I'm Nigerian and African-American. I graduated in 2021. I'm from the Bay Area, born and raised in Oakland, California. I majored in Environmental Studies with a concentration in Global Environmental Justice. In terms of campus organizations, I was a part of the Black Student Union where I worked with the Student Union Assembly (SUA), Office for Diversity, Equity, and Inclusion, Ethnic Resource Centers, Rachel Carson College, and others.

Selene: Hi, my name is Selene. I go by they/them pronouns and part of class of 2021. On my maternal side, I am a descendent from the Pascua Yaqui, Diné, Hopi, and Cherokee nations. I am originally from Temecula, California, and currently reside in Marietta. Like Marissa, I graduated from the ENVS department with a concentration in Global Environmental Justice. I was also a Political Science major. I was part of many student-led campus organizations.

Jordan: My name is Jordan. I like they/them pronouns. I am very mixed, on my mom's side I'm Punjabi. On my dad's side, I'm Swedish, Turkish, English, and a couple of other things. I'm starting my fifth year as a Ph.D. student on campus. I've been a part of the Diversity Committee in ENVS for way too many years, and the grad advisor for the Global Environmental Justice journal, which was pretty cool. I also started an official club for BIPOC in ENVS.

Kimie: Thank you to all for sharing those introductions. What was your background and previous experiences with the environment prior to UCSC?

Maxim: So born and raised in LA, my experience with the environment was limited to the beach, which I love dearly. It is still probably the only kind of outdoor experience that I would say that I actively enjoy. It was interesting majoring in Environmental Studies because I came here and everyone's like "let's go camping or hiking." I don't want to do that. I just want to go on my computer for the most part and look at cities. My experiences with ENVS are limited to the Environmental Science class I took in high school and going to the beach. I would also extend the environment to urban environments in which I'm a big city nerd.

Jordan: I'm from Boston, Massachusetts. I study food and agriculture. The whole reason I got interested in Environmental Studies was because there are really severe eating disorders among the women on both sides of my family. It was my own process of struggling with the eating disorder as a teenager and then coming to work on a farm as a healing process and thinking, "Why would I do anything else, but be on farms?" Also, UCSC is really famous for its on-campus farm, so that's why I ended up here.

Selene: I'm like Maxim. I am a city kid, through and through. My relationship with the environment was kind of non-existent until I came to UCSC. I would sometimes joke that UCSC was similar to Camp Crystal Lake, being in nature and everything. At first, I was very resistant but came to love it. It was a matter of weeks for me to go from, "I don't know anything about the environment" to "I love everything that has to do with ecosystems, ecology, anything that has to do with humanitarian-based ecology where you're focusing on environmental justice." I really fell in love with the ENVS department at UCSC. The one thing that I focus my research on and have always been fascinated by is natural disasters. A lot of kids would be watching Saturday morning cartoons while I was watching documentaries about natural disasters. I didn't realize that it could turn into a field of study and research.

Marissa: My background and previous experience with the environment began with my family. My grandparents built for a nonprofit that's based in the Amazon rainforest working with Indigenous people, land rights, and sovereignty, trying to protect the sacred headwaters ecosystem in Ecuador. I grew up surrounded by a lot of folx that are doing work to bring awareness around environmental degradation and the impact it has on marginalized communities. I also lived in Ecuador for a bit back in 2008, where I was able to visit the Amazon rainforest and the Galapagos and look at the different ecosystems in a different part of the world that I wasn't familiar with. This had a significant impact on my relationship with the environment, not necessarily urban environments but particularly our ecosystems and trees in general. I didn't really know about deforestation until I started learning about the purposes of clear-cutting for us, such as corn crops or cattle grazing land. Then, in high school, I took AP Environmental Science which solidified that this is something I'm interested in. What I didn't anticipate is the field being so white. I think that was what led me to search for some other kind of supplemental space for learning. Having learned a little bit more about environmentalism and environmental justice, what I've definitely recognized is the need for it to be more centered around BIPOC experiences because those are the folx that it's directly impacting the most.

Mainstream Environmentalism

The next section highlights participants' sentiments surrounding mainstream environmentalism. The history behind mainstream environmentalism is characterized by "racist, sexist, and classist" ideals and priorities (Curnow & Helferty, 2018: 148), which may unfortunately be widespread within the campus community. Understanding how mainstream environmentalism may be manifested within UCSC and the corresponding critiques contribute to a more socially aware and progressive on-campus environmental movement.

Kimie: Thank you. We'll move on to the next section which is environmentalism more broadly. The questions are, "What comes to mind when you think of mainstream environmentalism?" and "What opinions do you have, if any, of mainstream environmentalism?"

Marissa: Honestly, carrying capacity. It was one of the introductory classes where we talked about that so much. I really did associate environmentalism with whiteness. My own journey to decolonize my relationship with the environment has been a personal one. It's because I've actively searched out authors and different resources to learn more about the history of Black people or Indigenous people in nature. It hasn't been through our structured education system. I also think of green capitalism and regulation. One of the big issues that I have with our EPA and other political entities that do environmental work is regulating as opposed to getting rid of the problem, but it's more complicated than getting rid of one thing.

Maxim: As mainstream goes, there's a large movement of people who are interested in maintaining a habitable planet. The other thing that also comes to mind is the constant, co-opting of whatever movement there is by bad faith interests, similar to what Marissa was saying. We see this through corruption, which is a bit of a strong word but it is part of the language we use to talk about these topics. This is stuff we see in the news, like natural disasters. To what degree should we even be referring to natural disasters as fully natural? Or the idea of carbon footprint and the simplification that the onus of our current situation is mostly shared by individuals rather than a very select group of institutions and organizations. Then we say "saving the planet," rather than saving each other in our society—this shifting of narratives by bad faith interests through the adoption of neoliberal jargon. It's something that I always think about when I read the newspaper, watch the news, etc. You always see these things and think, "why are we using these terms that are kind of manipulative and sneaky?" That's my general idea of mainstream environmentalism.

Jordan: This conversation gets me so fired up and it's so interesting to reflect on. I also think of carrying capacity when it comes to mainstream environmentalism, that's the whole book that Paul Ehrlich wrote about. Also, Indigenous politics are about the environment and are all land-based. Obviously, it's environmentalism, but somehow Indigenous politics goes into Indigenous Studies and not in the environmental movement. It's not considered to be part of the mainstream environmental movement at least, which makes me think, "okay, I guess the mainstream environmental movement's reserved for white people, and everything else goes into BIPOC movements." That doesn't really make sense, but it's common sense that we have that in the U.S. Also it's very neoliberal; it's done through market-based mechanisms (e.g., we just need to buy stuff that doesn't pull the environment as much or buy with a specific label). Honestly aside from being racist, classist, etc., it's just not effective.

Selene: I see mainstream environmentalism as a very narrowed and limited perspective and understanding. It doesn't have the holistic inclusivity of the Global South even though they are usually the ones who have some of their teachings stolen from to make it more palatable and teachable to folx… It comes with a filter, the sense of elitism surrounding it. Who is really benefiting from these narratives that are made for the environment? To me, mainstream environmentalism is all performative and doesn't hit the substantial benchmarks that it needs to hit in order to address what is really going on and who is benefiting from this misperceived conflict. We think it's because mainstream environmentalism narrates in one way that doesn't capture the whole picture, and then we have a persistent issue.

Maxim: I would say I love Selene's use of performative in the context of environmentalism. ExxonMobil tweeting out, "Our thoughts and prayers with the people of Louisiana" is peak performativity. Let's say there is an individual saying, "I use and buy the green products," "I bring my own straw," etc., they would probably call themselves environmentalists. Is that a positive contribution to the environmental movement? I don't know.

Selene: I want to add to Marissa's experiences with the courses as it was a very interesting experience. With all due respect, I love the professors, but I feel that there were some missed opportunities to explain how the founders or "pioneers" of environmentalism were very much perpetrators of white supremacy towards Indigenous people and just enabled the genocide of Indigenous people in North America. The main pioneer I'm alluding to is John Muir because that was glossed over, just how he viewed Indigenous people when finding modern-day national parks. That was something I had to find out myself. That's an example where you know some aspects of this department really look over certain important facts that can contribute to what not to do when idolizing aspects about environmentalism.

Marissa: This is a generalization but I feel like it has to do with our department upholding a narrative about classical environmental studies and classical environmental science that's very Eurocentric and white-centric. It doesn't really have space for other groups of people that maybe had a deeper connection to this term environmentalism that we've created and has very close neoliberal ties. I think that's really been the trend of the Environmental Studies department. I've had my most positive experiences of Environmental Studies courses not in the ENVS department, but in other departments taught by lecturers that weren't even tenured or adjunct professors, knew how to diversify the reading list, and included the voices of marginalized groups that should be at the center of this conversation instead of one module covered in week nine at the end of the course.

Envisioning Equity

As panelists reveal the downfalls of mainstream environmentalism, the conversation transitions into an ideal and equitable environmental field—one where all voices and perspectives are incorporated and celebrated.

Kimie: What does a diverse, equitable and inclusive environmental field look like to you?

Maxim: I'm going to twist the question. A major moment for me was when I had to read a book for [class]. The author posed a question of how are you supposed to have a positive impact on this environment when we cannot even handle our relationships with each other as people. How do we address this issue that does not necessarily fall within our five senses when we can't even resolve issues related to each other? Since COVID and the racial protests, issues of race, class, gender, and the same wealth inequality disparity have continued to grow. These are a bunch of situations in which there was a lack of empathy and no consideration for each other. When it comes to environments, how do we draw to that? I don't know. My personal interest with the environment is urban systems which are not exactly ecosystems. I like urban ecosystems because they represent these dense, mass nexuses of human interconnection. Clearly, we're demonstrating that we currently do not have that. We're currently failing to support the interconnectivity of each other within countries, states, cities, and municipal districts. As far as the question goes, what falls within the environmental field? I'm not exactly sure. I think that extends to a lot of positions that we probably wouldn't even be thinking of.

Jordan: I want to pick up on this idea of interconnectivity and relationships. The environmental studies that I want to embody is all about relationships where it's us in an Indigenous politics thinking where it's not about humans and the environment. It's about our relationships with each other and our non-human relatives. This way of developing relationships to me is everything as somebody that's not native to Turtle Island. My whole biological family lives on the East Coast. I'm out here by myself. How do I even feel like I belong here? That sort of process, especially with COVID feeling so isolated and not being close to family. I really feel like I have to build these relationships that Indigenous folx have been calling for all along. That will help me feel like I have a place and community here. If I have deep relationships with people and non-humans, I can't exploit anybody like the massive amount of violence that capitalism does. To me, it comes back to the politics of relationality and wanting to care for one another. To me, the root of a diverse, equitable and inclusive environmental field is what Maxim was saying. It's not what is an environmental field at that point.

Marissa: I echo what everyone has said. Jordan reminded me of a realization that I had in my sophomore year which is that you can't make money by having positive relationships with people. That's a problem when you're based in a society that's built around capitalism, especially in the United States where we live in a society that's centered around the individual and not around the family. In my personal family, you have to share, think about other people, and make compromises. But that's not how our society, at least in the

Western sense functions. It's very much focused on the individual. You can sell to an individual but it's hard to sell to a group of folx that have a symbiotic relationship with each other, where we do things for the benefit of the group and we're going to make compromises.

Maxim: Marissa blew my mind with the idea of needing to make money and positive or negative relationships because I'm thinking about how we have so much cool technology. People are trying to collectivize and organize on platforms like Twitter, Twitch, or Reddit, but then they get dominated by parasocial relationships.

Selene: I think what Marissa said was very important, especially with the connections part because it frames positive relationships and connections. Positivity comes from impressions, and it comes from who we interact with. Are you trying to make that relationship positive? When thinking about a diverse, equitable and inclusive environmental field, you have to think about who was excluded and why they are excluded. Who is struggling to prosper in the environmental field? What do we define as prosper? Do we mean someone who gets wealth in recognition? How do we define those? Those are questions that come to mind.

Maxim: That's great. I was thinking about what all three of you just said and it triggered a memory. I took a class on Marxist theory which was really inspiring. Towards the end of a Zoom class, the people were goofing off. The TA said "No. I'm going to stop you all there." He then discussed his relationship to students, his relationship to the institution, and the different relationships. He said, "I'm currently an employee of the institution of UC Santa Cruz, and right now you all are my students for this. That is our relationship. We need to be very clear about what the boundaries are." That was actually really important. As far as answering the question within the context of the undergraduates, being educated about your relationships with the people around you, the people who are mentoring or teaching you, and your relationship to the university that you're currently paying for is super important. I think that would be very beneficial to understand early on in the process, and not having a graduate student in a Marxist theory class having to explain that to you. I think this is critically important to know to be able to navigate most spaces in general, especially a space like the University of California.

Jordan: That's so cool that you bring that up, Maxim because I have thought about this too, with regard to the [2020 Cost of Living Adjustment] strike. Obviously, I was kind of involved in the strike because I'm a grad student, but it brought to my attention the fact that the environment of the university is not really considered a realm to talk about in environmental studies. It should be. Let's talk about our immediate environment like that room in the [Interdisciplinary Sciences Building] 221, where all of the classes are. Let's talk about where we are, and whose land we're on. Also, we're all sitting in these desks in front of somebody who's supposed to be the expert. How are we supposed to feel like we know anything with this arrangement of the classroom? I think that's so cool that your instructor took time to talk about that because it's like if we don't understand the environment that we're learning in, we're going to think that all this knowledge is just the truth, when it's totally filtered through the space in which we're learning it.

Maxim: As far as the strike goes, I think that's so important because I was there at the time of the strike as an undergraduate. There were a lot of people who were very confused about what was going on because they had never been near a strike. They didn't have any kind of education about their relationships to the land, people, and hierarchies of people around them. How are you supposed to understand what exactly a strike entails? Obviously, there's what we've been taught about, but being in it is very different. There were a lot of undergraduates whose hearts were in the right direction and joined but were still clueless about exactly what's going on around them, and having that kind of understanding about relationships would have been incredibly valuable.

Marissa: To add, I think it's by design, having worked extensively with our administration and different offices that you think communicate with each other, but don't. It's very clear that this lack of building a space where communication is at the center of building relationships is not really the priority of the UC system, as that has really been my experience. I think it begins with asking questions. Like Jordan said, the way our spaces are built within our education system is the expert and then the uneducated mass that has to listen to what the expert is saying and not necessarily ask questions or think critically about the information that's being shared. I think that's super great information to think about that isn't really considered. I haven't really come across that as a conversation in any of my other undergrad spaces, it's usually just outside, clubs, or general groups of friends. It's not really like within the curriculum to talk about these spaces.

Selene: I have two points to bring up. Marissa and Maxim both brought up one and Jordan brought up the second point. With the uneducated masses dynamic, it reminds me of the banking concept, the dynamic between an expert and learners. Whatever the expert says, the students process it and don't think beyond what it's taught, just regurgitating what is said back to the expert. I think that is what exacerbates this miseducation that allows this course to happen in critical events like the strike.

The second point is about Jordan's land acknowledgments. I have such a hard time believing certain academic departments are all about Indigenous advocacy and justice, because they will talk about Indigenous people and what they go through without actually having them in the room, and to participate in these discussions, or to even know what is the first thing to do in order to acknowledge the land and our surrounding environment. We have to start thinking about the immediate environment of UCSC like, "Earth Science involves natural elements. Our campus is on a limestone mine and this city on a hill is a coastal town." When we're thinking about our immediate environment and what we pay attention to, we have to understand ignorance and intentional disinterest. Educating yourself is intentional. When it comes to the strike and environmentalism, you really have to take comfort in the fact that there may be some ignorance. That's warranted and benefit of the doubt because it provides teaching outside of this banking concept of just processing information as is, but also understanding who was intentionally disinterested in what is being said and what are we expanding these discussions on.

Marissa: The whole intentional disinterest is so frustrating because education is about choosing to educate yourself about something. You can't force someone to choose to be interested in something, they have to choose that interest themselves. It helps if their shared identity overlaps with whatever they're interested in. In my case, I experienced racism, sexism, and Islamophobia. Those things are interesting to me and I'm going to educate myself about how they interact with other aspects of life. But if you're a cis-, het-, white, straight man who lives in upper middle-class suburbia then maybe those identities aren't that important to you because they don't relate to your lived experience. It jumps back to performativity, like how Santa Cruz is super environmentally friendly or has this narrative of being a really social justice woke liberal college. I would say it's very performative. A lot of them are rooted in neoliberal and colonial ideals that are outdated, especially within the ENVS department. I think there's a lot of hesitancy to go "radical" and leave this well-worn path of what classical environmental justice is or these overdone narratives about environmental justice. One should reach out to the folx, like Selene said, that are directly related to those issues, whether they're Indigenous people or BIPOC, and have them in the room. Have those people lead the conversation as opposed to people who don't have any of those lived or shared experiences leading the conversation who often do a poor job at it.

Sustainability on Campus

After sharing their thoughts on mainstream environmentalism and an equitable environmental field, the remaining scope of the discussion focuses on life at UCSC, starting with participants' perspectives on sustainable ideals and its application at UCSC.

Kimie: Our next topic is sustainability on campus. What comes to mind when you think of the sustainability movement? What opinions do you have, if any, of the sustainability movement?

Marissa: I know about PoCSC, it's the People of Color Sustainability Collective. I remember the way that I learned about it was because I was interested. I wasn't exposed to their workshops and information sessions that much if I hadn't taken the initiative to go to the Ethnic Resource Centers and asked around about how people of color are getting involved in sustainability on campus. I was pleasantly surprised to find that there was a space that was working towards bringing more diversity into the conversation of sustainability. In terms of the larger topic of sustainability, I think there is a mindset that's in Santa Cruz. There is a narrative of a certain type of sustainability of being better. For example, growing up, my family would reuse old peanut butter jars or jars of food that we finished by washing it out, and then it would become a makeshift mason jar as a way of practicing sustainability. But when I came to Santa Cruz, it was very much like Hydro Flasks which I already have, but they're expensive. Also, when you go to purchase Birkenstocks and other sustainability products, they're really expensive and not that accessible to folx who might not have the budget to just drop $50 on a water bottle. I thought it was an interesting way to be sustainable. I felt that there was a certain way to be sustainable at Santa Cruz that was mainstream and streamlined that didn't necessarily include other ways of being sustainable.

Maxim: To push on that thought in terms of viewing sustainability, I think there are two kinds of versions I saw. There's the individualized sense of sustainability which is to not diminish, but these market-oriented approaches of buying this water bottle, shoes, etc. Then there's the greater topic of sustainability. I remember one of the major movements was Fossil Free UCSC, a very different type of sustainability. Students would be like, "I want to learn how to be sustainable in my individual life," which is through buying these products or learning how to grow my own food. Then there's this kind that is more societal or on an institutional level, so Fossil Free would be one of those greater forms of sustainability. I feel like there are those individual kinds of sustainability movements when you have all these different programs, groups, the Sustainability minor, PoCSC, CALPIRG, and then institutional sustainability movements which are very distinct. As far as their presence on campus, I wasn't really sure as I did not engage in them too much. I'm curious if you had observations about that.

Jordan: I felt those two that you're talking about Maxim, and then there's a different kind of sustainability that emerged. It seems to have emerged not from people being interested in environmentalism, but a roundabout way. The fact that PoCSC is housed in the Ethnic Resource Centers is really interesting because it's not people interested in environmentalism that end up finding that. It's more folx interested in being part of the Ethnic Resource Centers, and then they're like, "Oh, I actually am interested in environmentalism too but I never thought it could start from concern for BIPOC communities. I thought I had to start with environmentalism." Then we have this roundabout way of critiquing it. It's interesting because being in ENVS, I think we're exposed a lot to those first two kinds. But then there are people in Critical Race & Ethnic Studies (CRES), or in Feminist Studies, who are all fired up about environmentalism or I'm just like, "Oh, I had no idea that you people even talked about environmentalism. I thought you study gender and race." I have met people on campus that are talking about environmentalism in a really different way, but not through the ENVS program.

Marissa: I want to add. I don't want to shade the ENVS department too much. But to be honest, a lot of my most interesting and critical conversations about the environment have been with people of color who weren't necessarily in the ENVS department, but rather in my CRES, Sociology or Anthropology courses with other students of color. Like Jordan said, the conversation would go into food deserts, urban sprawl, or redlining, then transitioning into environmental justice issues. Personally, I've really enjoyed those conversations more than facilitated conversations in the ENVS classrooms that are being posed by the professors where the conversation feels a little bit structured. The question was framed in a way to lead us to a certain conclusion about what we're learning in class that day, as opposed to it being a little freeform. In other words, letting the students think more critically about their own identities and how it relates to the course matter and letting whatever conclusion that comes from that conversation be the end result as opposed to being funneled into a certain way of thinking/looking about the environment, environmental justice, or environmental studies. So I thought that was interesting as well.

Selene: I think the problem with the sustainability movement at UCSC is that students assume that just because they are part of that institution means they're automatically an environmentalist without having to do the mental and physical work to actually meet the sustainable efforts that are being put forth. UCSC definitely has a stigma around it where we have the Hydro Flask and Birkenstocks, like Marissa mentioned, and if we look the look, then we're a part of that. Actions speak louder than words.

The UCSC Environmental Studies Department

As described by the participants, sustainability on campus is a critical area for potential improvement. The following topic is the panelists' views on racial composition and its influence on the departmental climate, revealing another area that can be further developed.

Kimie: This question is, "ENVS at UCSC is and has been predominantly white at the undergraduate, graduate, and faculty levels, what are your thoughts on this and how does this make you feel?"

Jordan: I think within ENVS, it's so obvious that the department is not for students of color that it's almost not worth talking about to me. But I think what's interesting is that so many students of color get an environmental education on campus, without that coming from the department because there is so much student organizing and there's so much interest, no matter what your major is and no matter what your residential college is. I think this is one of those places where I see myself and other students just leaning into other communities and systems of support outside of the department because it would be so much of a lift to try to make this department a supportive community that it's not worth any of our time. Certainly not as students, and maybe I see faculty members willing to put in more of that work, but having a community outside of the department is what makes it livable. Basically, I talk to two people in the department, two students kind of on a regular basis and then nobody else. So that's kind of how it makes me feel, avoidant.

Maxim: Like Jordan said, it's not that many people from ENVS. You have to lean into the other communities. For me, I couldn't really, because I spent literally all my time in the department as opposed to my work community, which was a little POC hub. There's a sense of exhaustion. I would watch cohorts of each undergraduate class come through and I'd advised them. It was the undergraduate advisor who said, "Maxim, please don't make yourself a totem of color for the department, because that would be hiding what is actually going on." As the years went by, it was exhausting when you see waves and waves of students come in and be overwhelmingly white. Then I'd go from those classes to my own classes and there was this sense of exhaustion to the point where it's like, "yeah that's why I didn't make a ton of anything more than collegial relationships with students beyond that." I found solidarity amongst my friends or other communities outside of the department.

Marissa: When Selene was speaking, I thought about our internship office which has the most internships offered or one of the highest numbers of internships that are offered to students for extra outside of the classroom opportunities to learn and supplement your academic experience. [Maxim writes in the chat: "very specific genre of internships."] I remember thinking, "wow, that's so cool," but then being really disappointed when I was looking at the listings of internships that were offered. I thought, "Wait, how is this the most that the department has to offer?" Notably, there's an organization called "Food What?!" that works with Black and Brown kids and getting them into farming. I thought, "wow, I can fight with them. I'm not super into agroecology but because they're a person of color specifically targeting marginalized communities, maybe that's what I'll do for my senior internship" which I didn't end up doing.

On a personal level, I echo everything that Maxim and Jordan were saying because it was very apparent from my freshman year. Going into the classes that had 100+ students, I was the only Black student in my class and sometimes the only female-presenting person in the classroom. That was very visible so, it was obviously isolating. Also, I was in Rachel Carson College which is super white. I picked Rachel Carson and thought, "Oh, this is going to be where people who are intentionally going into environmental justice are going to be." I didn't find those people in Rachel Carson or in ENVS. I found them in the CRES department, the Black Student Union, and my counter spaces that are created on campus as a sense of community, not in the actual spaces that were created by the department. The tone that it set for my relationship with the ENVS department was a sense of apathy for any of the resources that were offered. It was so whitewashed and alienating that I didn't want to spend more time in a white space longer than I needed to. I felt a sense of apathy and that's not my job to fix it, but I definitely recognized it as a problem.

Selene: I echo all the things that Maxim, Marissa, and Jordan said. For Indigenous folx, it is really weird when your culture is talked about so much from the lens of a white professor who has considerably grown up as very privileged and with a limited scope of perspective. Anything that challenges their perspective of how they view Indigenous people in the environmentalism movement is a little bit frowned upon. For example, I started getting really involved with Indigenous activism. I started translating that to my own projects which backfired because my own type of academic research was not necessarily celebrated in comparison to non-critical assignments made by students that would love to talk about Indigenous people in a way that it makes it seem as though Indigenous people are so distant rather than sitting right next to them in the classroom. I find it very difficult because if we're talking about ENVS, we also have to look at what's being celebrated. You can probably sense animosity in my tone. It's just weird for our department to talk so much about Indigenous activism when there isn't that level of prestige given to students who dedicate their work to BIPOC efforts and environmental activism. To me, that's how you know there is a specific type of nepotism and hierarchy.

Maxim: To follow up, Selene made me think of something similar to what you just described, specifically the part about who and what is celebrated within the department. I have pretty thick skin. The only time that I felt real genuine anger within the department was a professor who everyone would worship. This [law] class is where I think there were only two other people of color in the class. One time in class, this guy flexed his boat and how awesome he went for a boat ride. I was thinking, "why are you talking about this?" The sad part is I think the class and the structure of the class were interesting and I learned a lot. But I felt so alienated because the way he composed himself in the class is this kind of machismo, teaching this content I think was really important but also catered to a specific audience. That was one time I experienced the overwhelming whiteness of the department that got to me. I think law is something that's already inherently catering generally to white folx. It's related to money, getting into law enforcement, who's policing, what are they policing, etc. It's generally some major system of oppression.

Moving Forward

In the final section of the roundtable, the panelists give concluding remarks to share their own reflections from this conversation and impart knowledge on improving environmental spaces for BIPOC students.

Kimie: To wrap this up, could everyone have a closing statement?

Marissa: Honestly, I'm conflicted with having some sort of nice clean closing statement because I think it is pretty messy and problematic. I kind of like the idea of leaving this space of discomfort for whoever's going to be reading this roundtable discussion because the sense of discomfort that I felt the past 4 years in ENVS was overwhelming and I constantly had to deal with tiptoeing around other people's feelings or making other folx feel comfortable with my identity as opposed to the other way around. I guess the sentiment that I'll end with is that there's a lot of work to be done. A lot of the work needs to be done by white folx in the department. It's not necessarily the people of color that need to do the work because we're already doing it whether you see it or not. If this conversation made you feel uncomfortable, then there's a lot of things that you kind of need to reevaluate about your positionality and identity in the department. There's a lot of work that needs to be done internally to then reflect on making the department a more safe and welcoming space for students of color on campus and have a more authentic representation of what environmental justice really is.

Jordan: I agree with that. It can't be wrapped up. If it could be wrapped up, everything would be fine. But it's not. Part of a closing statement for me is who am I talking to. If I'm talking to other students of color in Environmental Studies, I feel like there's a couple of different things to be done. I think community-building is awesome. I think writing to each other, talking to each other, hanging out, growing food together, getting land back, and all this stuff is stuff that I want to do together. When it's talking to white people or people who are in positions of power in the institution, it's like "why are you so oblivious? It's not that hard to see this stuff." Kimie and I had talked at one point about putting together a 'zine of alums or students of color in general in this department to share experiences, but then who would consume this sort of sharing stories together? It feels really good. I get fired up because I feel like we can all share these stories and there's at least some way in which we're on a similar page about it. But then I don't want to share it just so that white people can consume that and then use it as data to figure out what to do about it. If you don't know, you're not going to know. If you know, then you're already doing something about it. So, I don't want it to be consumable either.

Maxim: As far as what I want, the logical start is to get more faculty and staff who are not just white dudes and diversifying that. I was very readily aware of the roadblocks that already exist like hiring more [people of color]. I wish there was a better avenue for students to voice exactly the things that we're voicing, ideally with protection or anonymity. It always seemed like this would be a really good start. It can't just be a few faculty and the undergraduate advisors who are supposed to be the champions of diversity and equity within a department. That's way too much. That's just not their job. That's a huge burden to put on a few people.

Selene: It's so difficult putting on different hats while navigating these departments that are predominantly white just because you are not only a student but also the educator for anything that's a common narrative for a lot of students of color. I have so much frustration. These are supposed to be our tenured educators, yet they're not necessarily doing the education that captures the range of topics that students are interested in. The only thing that keeps me going is knowing the fact that I have the potential to cause destruction and discomfort in the institution, which is obviously not the softest words to say but they are necessary measures to make sure that those who have backgrounds that come after me are able to navigate the institution easier because of what I or my colleagues around me put forward. For a lot of us, we're walking around these institutions not even realizing that we're breaking glass ceilings or carrying gold medals of identity around our necks just by being there. One thing I found interesting is that we touched base on was the fact that we're having this discussion. We all were educated by the ENVS department through our experiences or the curriculum. I think it was very interesting learning how we all got here, sharing potential moves for where these discussions are able to show progression, and showing the ability of a department to really consider a harder look at itself critically.

Kimie: Thank you, everyone, that was a lot. I hope that this book chapter is just the start of the conversation for a lot of folx in higher education spaces. I hope you know that you are being heard. Thank you all for coming and I hope this was a good experience for you all.

Discussion

The conversation from these selected individuals of color expresses the utmost importance of diversity, equity and inclusion within environmental sustainability spaces at UCSC. Embedded in a redwood forest with scenic flora and fauna, UCSC attracts many sustainability-focused individuals whose ideals can often be categorized as mainstream environmentalism. From posting blanket Twitter statements to buying trendy sustainable products, these well-intentioned but performative and minimal actions can have a counterproductive effect on promoting inclusive sustainability, emphasizing the significance of genuine efforts that incorporate the diverse perspectives and experiences of people of color.

Kimie's Reflection

The articulations from all contributors, including Riri and myself, demonstrate the various intersectional identities that students bring to ENVS. In turn, participants are drawn to pursue the transdisciplinary interests that the department has to offer, ranging from agriculture to urbanism. Despite the diversity in interests and backgrounds that are showcased, everyone's sentiments reveal that ENVS is not conducive for the well-being of students who come from such diverse and underrepresented backgrounds.

Upon analyzing the discussion, I noticed that the participants' sentiments parallel my own experiences. The panelists share numerous experiences from not finding a community in ENVS to being disappointed in the quality and quantity of which their culture is represented within the curricula. As a woman of color undergraduate in this department, I find myself remembering similar feelings of isolation in my early undergraduate days, which could explain why I was attached to the few mentors who did reflect my background. While I am appreciative that I got to experience firsthand the true impact that faculty diversity can have on an undergraduate student's socioemotional and professional development, I know that my situation is unique and that unfortunately, many students in this department do not have the same experience, indicating the need to prioritize the department's faculty diversity.

As the facilitator of this panel, what also resonated deeply with me were the tones and manners in which the participants spoke, two elements that cannot be seen through this transcript. The panelists communicated with much passion and frustration, sharing suppressed grievances and wishful knowing that the department can do better. Head nodding, laughter, and an active behind-the-scenes Zoom chat further reinforced their collective feelings and mutual experiences. Correspondingly, my own senior thesis was fueled by profound helplessness after experiencing a troubling racial incident in the department. I felt the same passion and frustration as these roundtable participants during the initial stages of my senior thesis. I understand how disheartening it is to not feel heard, as well as how empowering and validating it feels when someone does listen and to have one's thoughts permeate their consciousness. Similar to what Jordan and Maxim said, I echo that there needs to be more channels for student feedback, especially for BIPOC students whose voices are often silenced and unheard.

This chapter highlights a total of six BIPOC voices, but our sentiments are not in isolation from other BIPOC students in ENVS. These themes are similarly represented in the data I collected for my senior thesis. The additional voices, spanning a decade from the time they graduated from UCSC, shared similar concerns, such as experiencing microaggressions, facing discomfort in classroom discussions on racial and ethnic topics, and feeling disconnected from the different environmental interests as their peers. What this shows is that the concerns these students raise—feelings of isolation, tokenism, exclusion—are not novel or new, but part of a longstanding pattern of marginalization of race, gender, socioeconomic identities and other forms of social difference within environmental fields. Knowing that there were cohorts of BIPOC students who came before me and faced similar sentiments and that no significant changes were made to improve the racial climate makes me feel unsettled. More research on this topic needs to be conducted in order to improve belonging for BIPOC students in environmental and sustainability-focused fields. I also urge practitioners to be more inclusive with their pedagogical practices and provide guidance to BIPOC individuals who seek that support. Through these efforts, more BIPOC environmentalists may be retained and encouraged to pursue environmental activism which will be beneficial for environmental spaces, both in academia and more generally.

Riri's Reflection

Reading through this conversation, I am finding myself resonating with many reflections from our participants as I look back on my experiences as a first-generation woman of color from the department during 2014 through 2018 as an undergraduate and as a staff member from 2019 through 2021. I think about the time during my sophomore year in one of my required Environmental Studies courses, where I felt the courage to go up to the first faculty guest speaker of color in the department. The presentation she gave and the conversation I had with her, expanded my understanding of what interdisciplinary studies meant. It allowed me to understand how Environmental Studies was not only a place to understand how we can be sustainable and prevent climate change, but is also a field of study that also includes daily, lived experiences of our own students and environmental injustices these students in marginalized communities face.

The way students in this roundtable discussion talk about when and why they began to bridge the gap between themselves and their education roughly follows the many theories of student development that are foundations to higher education and student affairs work. I attended many professional development and adviser training sessions at UCSC, and these student development theories, first coined in the 1960s, heavily contributed to our discussions and often framed our training. These theories describe the psychological and social stages of development among college students. What is lacking, though, is a continuous engagement with critical race theory. As aforementioned and reflected in the roundtable discussion, students of color and first-generation students face many obstacles and challenges that inform many of their decisions throughout their college life. These theories reveal a limitation, especially in language around race and considerations for the roles of racism in students' development and learning (Patton et al., 2007: 39). As Patton et al. (2007) recommend, "embracing the critical race theoretical perspective is an important step in creating spaces for safe dialogue, reducing microaggressions on campus, and moving one step further toward understanding the intricacies of multiple identities, including race" (47). My recommendation and own reflection to the roundtable is that it urges us to reflect on how often racial perspectives are engaged into assignments, readings, and class discussions. Classrooms and advising spaces should be mindful of the roles that "race, power, and privilege play in classroom dynamics, particularly in predominantly White settings, where few students of color are represented" (Patton et al., 2007: 49). As the themes from the roundtable discussion show, students feel that this kind of mindfulness is not being practiced in the ENVS department, which is pushing them to find opportunities to engage with environmental issues they are concerned about outside of this department.

Currently, I work as a first-time freshmen academic counselor at San Francisco State University, where my caseload serves mostly students of color, low-income, and first-generation college students. Having conversations with students coming out of the pandemic during my counseling sessions, I am noticing an increase in the troubling pattern of social anxiety exacerbated by the COVID-19 pandemic and remote learning. I find myself struggling to instill self-advocacy in their education and the desire to claim their education rather than passively receiving it. More than ever, I rely on finding faculty and staff with backgrounds that these students can see themselves in, in hopes of increasing their motivation and ultimately their retention. With this lens, I am working to reevaluate how the roles of race and racism intersect the experiences of my students and recognize how it influences their decisions and interactions, and I urge other settings of higher education and student affairs across all disciplinary fields and institutional sectors to continue to engage in these perspectives to better support our students of color.

Final Thoughts

This roundtable discussion reveals that the immediate learning environment—from interactions with fellow classmates to course readings in syllabi—does influence BIPOC students' perceptions and belonging in both positive and negative ways. Unfortunately, environmentalism persists as a predominantly white field, making it difficult to dismantle the institutional and societal barriers inhibiting BIPOC students' success in navigating their spaces. To facilitate progress, the roundtable participants provide key ideas such as reevaluating hiring and retention practices for people of color; encouraging white folx to put more effort in educating themselves to minimize cultural taxation on their counterparts; and championing BIPOC voices within environmental communities as a means of empowerment to transform the field.

Beyond these actions as advised by the panelists, the paradigm within the higher education setting needs to be re-examined for a necessary culture change. Although the student development theory provides a strong basis for knowledge, expertise, and practice and serves as a foundation for many spaces on college campuses like ENVS, this type of "sense making" tends to only provide a general overarching perspective about a certain trend. This theory in particular, that was first introduced in the 1950s and 1960s, does not reflect the ever-changing educational landscape. Therefore, as graduates and former members of ENVS, we urge universities to continuously engage in critical examinations that draw upon critical race theory and evolve with the needs of the growing population of first-generation and students of color.

References

Adamson, J., Stein, R., Ortiz, S., Leal, T., Peña, D., & Dixon, T. (2000). Environmental justice: A roundtable discussion. *Interdisciplinary Studies in Literature and Environment, 7*(2), 155–170. http://www.jstor.org/stable/44085765

American Council on Education. (2019). *Enrollment in undergraduate education.* Retrieved October 15, 2022, from https://www.equityinhighered.org/indicators/enrollment-in-undergraduate-education/race-and-ethnicity-of-u-s-undergraduates/#:~:text=Undergraduate%20Education&text=In%202015%E2%80%9316%2C%20approximately%2045,of%20Hispanic%20or%20Latino%20enrollment

Bristol, T. J., & Martin-Fernandez, J. (2019). The added value of Latinx and black teachers for Latinx and black students: Implications for policy. *Policy Insights from the Behavioral and Brain Sciences, 6*(2), 147–153. https://doi.org/10.1177/2372732219862573

Carrell, S. E., Page, M., & West, J. E. (2010). Sex and science: how professor gender perpetuates the gender gap. *Quarterly Journal of Economics, 125*(3), 1101–1144. https://doi.org/10.1162/qjec.2010.125.3.1101

Causey, J., Kim, H., Ryu, M., Scheetz, A., & Shapiro, D. (2022). *Some college, no credential student outcomes, annual progress report – Academic year 2020/21.* National Student Clearinghouse Research Center. Retrieved October 15, 2022, from https://nscresearchcenter.org/wp-content/uploads/SCNCReportMay2022.pdf

Curnow, J., & Helferty, A. (2018). Contradictions of solidarity: Whiteness, settler coloniality and the mainstream environmental movement. *Environment and Society, 9*(1), 145–163. https://doi.org/10.3167/ares.2018.090110

Dare, K. (2021). *Propelling UCSC undergraduates of color in environmental studies through mentorship* (Senior thesis). https://transform.ucsc.edu/building-belonging-fellows-2020-2021/

Dare, K. (2022). *Propelling UCSC undergraduates of color in environmental studies through mentorship* (p. 2). Global Environmental Justice Observatory. Retrieved June 1, 2022, from https://globalenvironmentaljustice.sites.ucsc.edu/

Garcia, S. E. (2020, June 15). *Where did BIPOC come from?* The New York Times. https://www.nytimes.com/article/what-is-bipoc.html

Gardner, A. (2022). *Persistence and retention fall 2020 beginning postsecondary student cohort.* National Student Clearinghouse Research Center. Retrieved October 15, 2022, from https://nscresearchcenter.org/wp-content/uploads/PersistenceRetention2022.pdf

Grim, J. K., Bausch, E., & Lonn, S. (2022). The real-time social and academic adaptations of first-generation college students during the global pandemic. *American Behavioral Scientist, 000276422211182.* https://doi.org/10.1177/00027642221118260

Johnson, R. C. (2020). *Segregation in higher education and unequal paths to college completion: Implications for policy and research.* American Council on Education. Retrieved October 15, 2022, from http://www.equityinhighered.org/wp-content/uploads/2020/11/Segregation-in-Higher-Education_essay_individual.pdf

Kirp, D. (2019). *The college dropout scandal.* Oxford University Press.

Office of Institutional Research, Assessment and Policy Studies, UCSC. (2021). *2021 Fall quarter all undergraduate declared and proposed major headcount.* Retrieved October 15, 2022, from https://mediafiles.ucsc.edu/iraps/student-majors/fall-term/2021-22/fall-ug-majors-declaredandprop-mc.pdf

Patton, L. D., McEwen, M., Rendón, L., & Howard-Hamilton, M. F. (2007). Critical race perspectives on theory in student affairs. *New Directions for Student Services, 120*, 39–53. https://doi.org/10.1002/SS.256

Price, J. (2010). The effect of instructor race and gender on student persistence in STEM fields. *Economics of Education Review, 29*(6), 901–910. https://doi.org/10.1016/j.econedurev.2010.07.009

Whitley, S. E., Benson, G., & Wesaw, A. (2019). *First-generation student success: A landscape analysis of programs and services at four-year institutions.* Student Affairs Administrators in Higher Education. Retrieved October 15, 2022, from https://www.luminafoundation.org/wp-content/uploads/2019/03/first-gen-student-success.pdf

Kimberly Dare is a graduate from the University of California, Santa Cruz with a B.A. in Environmental Studies and a minor in Education. Her award-winning senior thesis focused on amplifying the voices and concerns of BIPOC (Black, Indigenous, and People of Color) students within the Environmental Studies department. She currently works in Environmental Health and Safety in the biotechnology industry.

Riri Shibata is a graduate from both the Environmental Studies and Legal Studies departments at the University of California, Santa Cruz. She wrote two, award-winning theses focusing on the daily, lived experiences of the indigenous Okinawan community under U.S. military occupation. After graduating from UC Santa Cruz, Riri spent her time as a Research Analyst for a social science research firm, mainly supporting K-12 education research in Santa Cruz County. In 2019, she transitioned roles and became the Undergraduate Program Coordinator and Academic Adviser for the Environmental Studies department at UCSC. Currently, she works at San Francisco State University as an academic counselor supporting first-generation, low-income, and students of color.

3 Teaching Critical Sustainability Studies: Towards a Relational Pedagogy

Emily Murai

Abstract

How do we teach disciplinary content, like American environmental history, that have exclusionary, extractive roots to underrepresented students who face their own marginalization in society and at the university? What pedagogical practices best empower students to make meaningful and substantive change within these fields? To begin addressing these questions, I explore the intersection between relational pedagogies, a humanistic pedagogy that places primacy on caring relationships to enhance learning, and critical perspectives within environmentalism. Used in conjunction, they create productive synthesis between pedagogy and content that enables students to work towards greater inclusivity in environmental knowledge and praxis outside the classroom. In particular, the concept of a "relational ontology"—a worldview in which "beingness" or one's sense of self is composed from relations with others—that undergirds relational pedagogies strengthens ideas of sustainability that go beyond techno-scientific or managerial approaches. At the same time, critical environmentalism pushes relational pedagogical practices to engage more deeply with questions of power, social justice and human-environment relations. I reflect upon my own motivations and experiences in using relational teaching strategies and offer suggestions around designing and implementing a relational pedagogical approach when teaching environmental courses from critical lenses.

Keywords

Teaching · Relational pedagogy · Critical sustainability studies · Care · Inequality · Humanization · Higher education

Introduction

Critical environmental scholarship, including work done by political ecologists, environmental justice advocates, and nature writers and artists of color have long called upon the mainstream environmental movement to broaden its purview beyond its traditional confines of environmental conservation and management (Pepper et al., 2003). Spurred by Black Lives Matter and other movements for social justice, some of these institutions are finally beginning to take notice. In the *New Yorker*, Jedidah Purdy (2015) provides an overview of the histories of exclusion and institutionalized racism that underlie the foundations of U.S. mainstream environmentalism. He details

E. Murai (✉)
University of California, Santa Cruz, CA, USA
e-mail: emurai1@ucsc.edu

F. Lu, E. Murai (eds.), *Critical Campus Sustainabilities*, Sustainable Development Goals Series,
https://doi.org/10.1007/978-3-031-30929-8_3

the racist and eugenicist views of some of the American conservation movement's founding figures, some of whom are also major political figures, like Madison Grant and Theodore Roosevelt. Similar views held by widely hailed twentieth century conservationists John Muir and Gifford Pinchot are documented in their own writing (Claborn, 2014). Mainstream environmental organizations, like the Sierra Club and the Environmental Defense Fund are now publicly denouncing these views and working to re-envision the role of these legacies (Brune, 2020; Fears & Mufson, 2020).

While important, this work has a long road ahead in addressing and overcoming these institutionalized historical precedents. Left unaddressed are questions around the environmental needs of communities of color, many of whom are at the frontlines of environmental and climate degradation. Mainstream environmental organizations often position themselves as purveyors of environmental justice, yet grassroots environmental efforts in these communities and how they can be best supported become eclipsed in these debates.

One of the ways the mainstream environmental movement can redress its troubled histories is by reconsidering what gets defined and what counts as 'environmental.' Critical sustainability scholars, for instance, continually ask broad conceptual questions to trouble taken for granted ideas, such as, "what is to be sustained and what is not? And who gets to choose and who does not?" (Greenberg, 2013: 56). Doing so challenges highly settled, normative categories reflected in mainstream environmental and sustainability literatures and discourses (Sneddon, 2000). Instead, critical scholars push to open up ideas to acknowledge the ways knowledge is fluid, dynamic and changing: to "recogni[ze], celebrat[e] and validat[e] diverse understandings and expressions of sustainability," seeing it as "processual, symbolic/cultural, and affective" (Lu et al., 2018: 65–66). Such an approach incorporates a wider range of perspectives and lived experiences, attending to and celebrating the ways sustainability practices manifest themselves in various localized contexts.

One of the purposes of this book is to draw attention to the ways individuals and communities who are historically marginalized are working to enlarge ideas of sustainability within higher education. Other chapters point out the contradictions between UCSC as a campus leader in environmental and sustainability initiatives, with its institutional policies that continually reflect mainstream environmentalism's foci on conservation, preservation and techno-scientific solutions, while more equity and justice-centered approaches, like PoCSC, remain marginal. In this chapter, I bring the complexities of these conversations into another important site within higher education: that of teaching and pedagogy. The key questions I explore include how do we effectively teach students, especially those from historically underrepresented backgrounds, subjects in fields like environmental studies that have disciplinary foundations rooted in racialized, exclusionary and violent histories? How do we help these students avoid becoming passively socialized into these systems of knowledge, and instead begin to see the connections between their own marginalization and the marginalization of certain ways of knowing in these disciplines? How do we inspire students to become agents of change within their fields? Can certain pedagogical practices provide a means to transform disciplines to be truly inclusive and socially just?

These issues are incredibly complex and there are no easy answers. To begin to answering some of these questions for myself, I've experimented with different pedagogical practices and have come to appreciate and value concepts within "relational pedagogies," an approach to teaching that emphasizes caring relationships as a means to enhance teaching and learning, particularly for students from marginalized identities (Aspelin, 2014, 2020; Bingham & Sidorkin, 2004). In particular, one of the field's core concepts, the idea of a "relational ontology," illustrates some of the key concepts in critical environmentalism and sustainability studies in a real-world way and has helped me take the task of empowering students to transform disciplinary knowledge seriously. Relatedly, the focus on non-human relations within critical environmental fields pushes the bounds of relational pedagogies beyond human relationships and educational spaces. Taken together, these two seemingly disparate subfields create an educational framework for teaching in critical environmentalism that builds upon each other, providing a potentially transformative educational experience in which content and practice is aligned (Ferreira, 2017; Freire, 2007). In so doing, this framework enables students to challenge the hierarchies and exclusions within mainstream environmentalism and other forms of knowledge production in a substantive and meaningful way and avoid the performative or superficial nature of other "inclusive" teaching strategies.

In the first half of the chapter, I offer a conceptual exploration of relational pedagogies and show the value of this approach to teaching critical environmental courses and vice versa. In the second half, I offer more practical, applicable suggestions. I share my personal reflections and specific teaching strategies I successfully integrated in my courses. I draw from my over two decades of experience as a woman of color educator teaching mainly at large, public research institutions. UCSC in particular serves a majority of students of color but ideally these suggestions translate and are relatable to instructors in other contexts.

Relational Pedagogies and Critical Sustainability Studies

Though the importance of warm, caring relations between teachers and students has been long established in educational research (Noddings, 1992), the specific concept of relational pedagogies has recently garnered fuller attention in the wake of the field's "pedagogical turn" (Graff, 1994). Research in this area has examined the specific psycho-social processes that inform cognitive and intellectual development (Reichert & Hawley, 2013; Baker et al., 2008; Warren, 2014); improved learning outcomes, especially for students from marginalized identities and/or who face "at risk" factors (Jones & Crownover, 2016; Gross & Lo, 2018; Caine et al., 2022); and the application of this pedagogy in various educational contexts (Agne & Muller, 2019; Chika-James, 2020; Edwards & Richards, 2002). Concomitantly, the literature draws from and contributes to several fields, including metaphysical philosophy (Buber, 2002; Wildman, 2010); humanistic pedagogy (Noddings, 2005); teacher education (Reeves & Le Mare, 2017; Trauth-Nare et al., 2016) and the learning sciences (Darby, 2005; Prosser & Trigwell, 1999). In contrast to mainstream educational practices that emphasize knowledge acquisition and content delivery, this body of work suggests that teaching and learning is a set of deeply social, affective and relational processes that, when activated, can stimulate learning for all students (Klem & Connell, 2004; Wentzel, 2009).

The relational pedagogies literature places primacy on at least three different sets of relations in instructional contexts: between instructor(s) and students, among peers, and within the overall learning environment (e.g., classroom and institutional spaces), which are themselves networked within larger social and economic contexts that inform and shape the educational context (Hickey & Riddle, 2021). These zones of relationality set the stage for what Buber (2002) calls the "dialogic encounter"—a moment when each participant "turns [to the other] with the intention of establishing a living mutual relation between himself and them" (19). Buber (2002) describes the dialogic encounter as fostering "I-Thou" relations, where two individuals, a teacher and student for instance, enter into each other's lifeworlds and bridge the differences between them in an attempt to find mutual connection. He contrasts "I-Thou" relations to "I-it" relations that are more transactional and exchange-oriented, and may reinforce or re-entrench—instead of overcoming—differences between the two individuals (c.f. Freire, 2007). Teaching can occur in both instances according to Buber (2002), but deep learning takes place only in the former, where a relational bond is forged. It is the humanizing impact of this exchange, when participants in the learning space are authentically seen, accepted and experienced fully by the others, that the "participation, coordination, co-construction and transformation" of all participants is stimulated and the learning process is enhanced (Biesta, 2004: 17; Warren & Lessner, 2014). In this formulation, schools and teaching spaces are opportunities to foster genuine human interaction and develop relational skills in conjunction with the acquisition of academic content knowledge (Aspelin, 2020).

I find a relational pedagogical approach helpful when teaching critical environmental courses. Of greatest value to me is its conceptual basis: as hinted at above, relational pedagogies are premised upon a "relational ontology"—a defining worldview that emphasizes the interrelationality of all things. Humans (and non-human things) exist "in-relation" to each other, not as ontologically separate or isolated entities, as is posited in mainstream Western scientific ontologies. As Wildman (2010) puts it, "the basic contention of a relational ontology is simply that the relations between entities are ontologically more fundamental than the entities themselves" (55). A relational ontology, then, puts relations between things at the center of reality, the basis on which all else is understood. The broader contextual set of relations determines the "beingness" of things, not an *a priori* separateness and individuality, as is the modern, Western ontological basis. It is less of an ethic, act or choice, but is instead a defining world view.

Centering relationality and a sense of collectivity and care are central to our understanding of critical sustainabilities as we have asserted in this book. Indeed, most ideas of sustainability—even mainstream ones—are premised on some level of recognition of the interdependence of both humans and non-human (or more-than-human) worlds, and thus in varying degrees strive towards overcoming the isolation and objectification of Western mindsets. Yet I believe centering a relational *ontology* (not just the act of relating or an ethics of relationships) offers a great deal of leverage that can push the boundaries of the field towards promoting long-term sustainability without the risk of falling back on human-centered premises that even the most well-intentioned sustainability debates sometimes do (Sneddon, 2000). A relational ontology inherently prioritizes non-anthropocentric outcomes: if the category of "humans" is only so in relation to "non-human" things (or, as Jones and Crownover (2016: 19) write, "if the self is always a socially situated self"), then by definition, humans must act in ways that sustain or nourish the relationship with other humans and our non-human worlds. If human existence is predicated on its relationship to other entities (both human and non-human), the relationship is already front and center and does not depend on one's willingness to enter into or see the importance of those relations—they are already central to one's state of being. The act of sustainability in this formulation is inseparable from the worldview that underlies it: actions that harm one's relations ultimately harm themselves.

Different bodies of critical scholarship, like indigenous ontologies and epistemologies (Whyte, 2017; Kimmerer, 2013); Actor-Network Theory (Latour, 2007); hybrid geographies (Whatmore, 2002); and feminist epistemologies (Longino, 2017) have drawn similar conclusions. They each articulate a vision of networked relationality between humans and non-human things in which the "beingness" of humans and knowledge is predicated upon a relational understanding of the world. This approach transcends—instead of working within—the Cartesian, Western subject-object divide. If the task of critical sustainability studies is to open up and broaden the definition of what is considered sustainable, advocating for and working towards a stronger relational ontology illuminates a way forward that understands both being and acting in the world as two sides of the same coin.

By the same token, as an educator, I believe the non-human, environmental content focus within critical sustainabilities studies meshes well and can extend various foci within relational pedagogies literatures as well. As noted earlier, relational pedagogies' primary focus has been to demonstrate the importance of caring, bonded relationships to learning that attend to a holistic, "whole learner" understanding of students (Noddings, 1992). This tends to emphasize building human-centric relations over considerations of content and curriculum. Using a critical environmental framework, I suggest that an explicitly "critical" curriculum has the potential to provide a learning experience that is consistent with relational pedagogy's values around relationship building and holistic teaching while enabling new insights on the relationships between humans and non-human natures (Ferreira, 2017; Papenfuss, 2019). A "critical education" empowers students to ask questions about the world around them and seek their own answers to draw out connections between their lives and larger political and social processes (Freire, 2007). Using Freirean critical pedagogy has been shown to transform students' understandings of sustainability and nature, instead of reproducing overly narrow technical or institutionalized definitions (Ferreira, 2017; Papenfuss, 2019). Similarly, educational research that examines relational pedagogies in conjunction with a "critical" curriculum shows the ways this combination powerfully impacts students from marginalized groups, as it increases their awareness of social and institutionalized inequalities that shape their lived realities (Cammarota, 2006; Cammarota & Romero, 2006).

To be sure, this argument is not just about a critically-engaged curriculum, but about emphasizing environmental themes and human-nature relations within it. Following critical environmental scholars who have long argued that the ways non-human natures and environmental issues are framed reflect broader political, economic and social agendas (Foster, 2000; Peet & Watts, 2004), I wish to push critical education perspectives to center environmental concerns and issues. In contemporary society, when marginalized communities are increasingly being forced to contend with large-scale environmental and public health risks like global climate change, COVID-19, destructive wildfires, an unprecedented number of hurricanes and floods and other so-called "natural" disasters, environmental issues are deeply intertwined with larger social injustices around racial and economic inequalities (Battle, 2022). Examining the social and political processes that create these communities' conditions of environmental vulnerability can illuminate how environmental issues and contexts stem from and exacerbate institutionalized social inequalities (Pellow, 2016). This also shines light on the significance of one's own lived experiences, especially for those who identify along multiple axes of marginality, as a valuable starting point for theorizing nature and the environment. This is especially so, as black feminist nature theorists tell us, when one's experiences of nature or the environment do not fit mainstream narratives (Field Station Five, 2020). Starting with one's own experiences and definitions of nature not only contests the privilege and elitism of mainstream environmentalism, but also shows that overcoming human exploitation and oppression is intertwined with overcoming the objectification and exploitation of nature.

Taken together, the key concepts and practices of relational pedagogies and critical sustainability studies reinforce and strengthen each other and create a productive synthesis between pedagogical practice and course content. Instead of simply integrating a general set of inclusive teaching strategies into one's courses, drawing from a relational pedagogy—with a relational ontology at its core—when teaching critically-oriented environmental classes enables students gain both conceptual knowledge (to think through, for instance, how a relational pedagogy helps illustrate key concepts of care, collectivity and relationality as articulated in critical environmental literature) as well as specific practices and tools they can use to act upon these conceptual ideals outside the classroom (e.g., asking themselves what are the specific skills and practices by which we relate to one another, including non-human nature) to work towards a broader version of the world in which they would like to live. In this way, both relational pedagogies and critical environmentalist concepts challenge some of the dominant normative practices within our present-day systems of knowledge production, like ideas of nature that rely on exclusion and erasure of others, or exploitive, extractivist research methodologies. Working together, they resist the depersonalization and instrumentalization of education, academic work and nature, thus attracting students who may otherwise be turned off by these dominant practices. In the next section of the paper, I discuss my own experiences integrating relational pedagogies into my courses on critical environmental themes.

Reflections of Relational Pedagogies in Practice

As an instructor of critical environmentalism and sustainability themed courses, my own lived experiences as a woman who identifies as Asian-American Pacific Islander (AAPI) has driven my academic pursuits. On my father's side, I'm a fourth-generation Californian of Japanese descent. Both my paternal great-grandparents and grandparents were farmers, the latter interned in World War II along with their young children (Murai, 2022). On my mother's side, I'm a second generation Chinese American. My parents were first-generation college graduates of the California State University system, and as believers in public higher education, they pushed me and my siblings towards state schools for college. I chose UCSC—the UC that was the most geographically proximate campus to where I was raised in the suburban Bay Area, but culturally worlds apart—and this turned out to be one of the most fruitful decisions of my young life. At that time, as one of the youngest UCs, UCSC still prized itself as an alternative to the heavily bureaucratic flagship institutions, one that could offer more personalized living and learning environments. The university still issued in-depth narrative evaluations instead of letter grades, offered a variety of smaller seminar-style classes, and enrolled relatively low numbers of graduate students, which meant undergraduates had a lot of direct contact with professors. In these smaller learning communities, I thrived. Away from my community that defined education through grades, test scores and professional achievement, I felt free to explore and pursue my own interests. In the process, as many others who encounter these intellectual freedoms do, I embraced the life of the mind: I relished complex, seemingly contradictory ideas, rigorous analysis, debate. I loved studying theory, philosophy, ideas that were committed to understanding the human experience across different dimensions. I majored in Psychology and American Studies, the latter a now-defunct department that was a precursor to the Critical Race and Ethnic Studies (CRES) department. I learned about the history of Asian-Americans, and Asian-American women in particular, and my fragmented, disparate experiences as an AAPI woman finally started to become coherent to me.

One of the most influential experiences I had as an undergraduate was working as a writing assistant for the campus learning center, which at that time was funded by the Educational Opportunity Program (EOP), which served lower-income, first-generation students of color. Surrounded by a warm community of educators who were similarly dedicated to academic excellence, I began to understand the important role of writing in fostering learning and discovery, especially for students who had been underserved by their prior educational experiences. For two summers, I worked with EOP students on their writing in a month-long Summer Bridge program, then continued to tutor the same students as they moved into their first year at the university. From the relative longevity and intimacies of our work together, I observed these students grow, both personally and academically. Writing stimulated their intellectual curiosities and desires for greater inquiry. Writing also helped them find their own voices while adopting (and oftentimes pushing back against) the discourses of the university, out of which they could define their own purpose and presence—and move more confidently—within the spaces of the university.

The richness of my undergraduate education motivated me to go to graduate school in the critical social sciences, where I took on more focused questions around nature-society relations. As much as I loved studying these issues, I quickly realized that graduate school and the academic world did not—and were not intended to—extend the supportive, nurturing and empowering experiences I had as an undergraduate. I was unprepared to contend with the rigid world of academic hierarchies and power dynamics. Most isolating was that I did not see many critical scholars who shared my identity or lived experiences. I had a hard time finding people with whom I could relate. I found refuge in teaching, which I found to be more collaborative and engaging, and my background in tutoring writing as an undergraduate led to opportunities to teach my own writing courses. These experiences were illuminating. I knew early on that if I pursued a career in higher education, it would most likely be in teaching and in a place (both geographically and institutionally) where I would be surrounded by my community.

Thus, while teaching- and service-heavy positions are often seen within the academic hierarchy as secondary to publishing and research, pursuing a teaching path was for me very much a choice, one that has allowed me to give back to my community through supporting the educational advancement and social mobility of the next generation of college students—the same kinds of opportunities provided to my first-generation parents when they went to college. When I started teaching at UCSC in 2009, it felt like a dream job: I was teaching at the same institution where I felt most academically inspired and at a place and with students I understood and cared about deeply. And over the years, as I've taught a variety of courses in different departments, I've witnessed wave after wave of students from various backgrounds turning towards a critical education for the same reasons I did: because these perspectives legitimate and enable them to make sense out of their lived experiences, empower them to meaningfully engage with the world around them, and give them a sense of purpose and of radical possibilities for the future. Validating their desires and interests and supporting their educational journeys has been central to my mission as a teacher.

During my time at UCSC, I deepened my commitment to teaching writing, continuing to see writing as foundational to honing strong critical analytical and communicative skills. I have taught a wide-range of courses, from first-year writing and research seminars, to writing-across-the curriculum/in-the-disciplines (WAC/WID) courses and courses in the environmental humanities. I've had the privilege of relatively low enrollment caps (<30 students), and a great deal of autonomy in teaching my courses, which has allowed me to experiment with different pedagogical tools and styles. While I have consistently drawn from active learning and student-centered pedagogies, it is only more recently that I have felt a pressing need to teach environmental issues from an explicitly critical lens so as to affirm and support students who have historically been marginalized by this field. Over the years, I observed students of color struggling to find their place in traditional environmental studies courses, despite having a deep care for environmental issues. These students have faced continued marginalization and disempowerment by mainstream environmentalist curricula that have not taken their interests, desires and lived experiences into account and expected them to conform and adapt to their conditions of learning, as opposed to seeking to change them. They have a deep hunger for an education that meets their needs and enables them to address the challenges their communities are facing head-on. Educational institutions have an obligation to meet these needs as part of their mission as public institutions to educate, retain and support a diverse range of students. For me this means creating courses that are student-centered, participatory, and reflective of what students want to learn, not just what the institution or the disciplines dictate.

In my classes, I have experimented with relational pedagogical practices with some success. Course evaluations suggest that students have a full, expansive learning experience. Below I identify some relational practices I have found to be the most valuable. While some of them may seem on their surface run-of-the-mill "best teaching" practices, framing them through the lens of relational pedagogies and using them in critical environmentally-themed courses has the larger goal of working towards a relational ontology. Centering a relational ontology allows students to challenge extractivist, exclusionary and elitist systems of knowledge production. While I do find inclusive and student-centered teaching strategies valuable to a certain degree, they don't fundamentally question or challenge students' conditions of their exclusion on their own.

Create an Inclusive, Culturally-Affirming Classroom Environment From the start of the term, I invite students to bring their full selves, the richness of their backgrounds, identities and lived experiences, to the class. This acknowledges and affirms the full range of cultural assets students bring to the class and creates a classroom environment of support and acceptance, which is crucial for students of color and students from marginalized backgrounds. It also emphasizes that our relations with one another is a valuable site for learning and knowledge production. This is a particularly valuable resource at large public institutions like UCSC that enroll a diverse student body.

To set up this environment, I often start a new term with a "cultural competencies" activity in which each student (and I) map out our social identities (e.g., gender, race, ethnicity, religion). We then discuss some of the ways these factors have shaped our worldviews and life experiences. Increasing awareness of our own and others' social identities does a lot of important work: it is an icebreaker in which students to get to know each other beyond superficial facts; it centers discussion on social differences, and in so doing, provides an entry point into discussing (and challenging) the way social norms and hierarchies govern social interactions; it highlights the varied, intersectional backgrounds each class member brings to the class and allows us to see how we can learn from them. Setting up these conditions creates a safe space for students to share who they are and continue on in their educational journey from wherever they are.

Draw from and Expand Upon the Relational Capacities Students Bring to the Class Students bring rich relational histories with them to class, but prior educational experiences, particularly with the banking model of education, conditions them to devalue skills and knowledge in favor of receiving "expert" knowledge from the instructor (Freire, 2007). In my classes, I elevate and center these skills as an important element in learning – not only are we learning by being in relation with one another, but we learn about the value of relationality within environmental thought and work to improve our relational skills as a part of developing a critical practice. Refocusing attention on and refining skills students already possess leads students to then ask themselves what other skills and knowledge might they have that are traditionally undervalued in higher educational spaces, which is exactly the questions I want them to be asking.

To begin this work, at the beginning of the quarter, we spend time discussing our class's course expectations and what we hope to get out of the class. A specific set of guidelines based on respecting one another emerges: communicating as clearly as possible, being present and authentic, actively listening and disagreeing respectfully. The class is asked to commit to adhering to the guidelines for each class, which signals their commitment to the processes and practices of building a relational community in the class.

From this list, I transition to an active listening exercise. The exercise functions to deepen relationships and increase students' metacognitive awareness of the way they participate in social exchanges (and in writing classes, to teach how to summarize effectively). In pairs, students take turns responding to a low-stakes prompt (e.g., what do you hope to get out of this class? What was the most valuable learning experience you've had at this school so far?). Each partner is given an allotted amount of time to speak, usually about 1 min or so. In that time, the listening partner's job is to hear what their partner is saying—not just the words coming out of their partner's mouth, but for the emotions and deeper meaning behind what is being said. The listener's job is to learn how to be present and listen effectively and to take in the totality of everything that their partner is saying without responding. After the time is up, the listening partner then "says back" what they heard their partner say. They are asked to withhold judgment or analysis of the content and simply reflect back what they heard along with their sense of their partner's feelings about it. Only after the listening partner has reflected everything that the speaker said can the partners engage in a conversation about the content in the remaining time. Partners then switch roles so that each has a chance to speak and listen. At the end of the round, I ask students to reflect on what the experience of talking and listening is like for them—which role they are more comfortable in, what skills they might want to improve, what the interaction felt like. Given how much time we have, we may do a second or third round for comparison.

For many students, this activity is enlightening. It may be the first time they feel the simple power of presence from a peer or classmate. For some, it may be the first time they have listened closely to another person without judgment. Listeners often acknowledge how difficult it is to not interrupt or ask a question or to curb their natural impulse to automatically jump into a conversation. Others see how easily they slip into (or feel uncomfortable in) the talker role, and need to learn to strengthen their listening and summarizing skills. The benefit of this activity is that it slows down interpersonal interaction so that each pair can step back and analyze the interaction and foster a better understanding of themself and their partner(s).

As the quarter progresses, we continue developing these skills through different sets of activities: structured and unstructured class discussions, presentations, small group activities, writing assignments, peer-reviewing work and instructor-student conferences. Throughout these activities, I often ask students to reflect not only on their cognitive but also the relational processes to increase their awareness and self-understanding in the ways they relate to others and vice versa.

Be Transparent that I Am Using a Relational Pedagogy and Show the Intersections with Course Content I am upfront with students that I use a relational pedagogy in my classes. I structure most of my classes as seminars and I have written fostering relationships and engagement with others into course outcomes. While I believe these skills are valuable on their own, more importantly, I want students to see the intersection between pedagogy and course content in fostering new imaginaries and practices around human-environment relations.

In terms of course content, I often frame questions of sustainability and environmentalism as an attempt to address problems that stem from the Judeo-Christian underpinnings of the modern Western world which, later, combined with emergence of capitalism, abstract scientific knowledge and the modern liberal state, constructed social hierarchies that enabled the objectification and destruction of the natural world (including people and communities associated with nature). As we delve into discussing ever-increasingly complex social and environmental problems, students start connecting their own existence to the environmental destruction they see happening around them (what can I do as a member of this society to create meaningful, structural changes that improve the environment?). I push them to think about the relational work we are doing in this class and the ethics of respect and care, and how they might be able to meaningfully practice these skills outside the classroom, particularly in transforming mainstream environmental thought and praxis. At the same time, we may take on readings that emphasize a relational praxis like indigenous approaches to climate change or decolonial scientific methodologies that helps to hit home how relational practices can be used outside the classroom in service of larger forms of environmental action or advocacy. Through this process, students begin to see core critical sustainabilities' concepts (e.g., inclusivity, relationality and human-non-human relationships) are not mere abstractions, but principles that can be enacted outside the class as well.

Integrate Course Materials that Reflect Student Backgrounds, Identities and Interests Course materials, like pedagogical practices, reflect the larger values and ideals of a class. Since the larger aim of critical sustainabilities and critical environmentalisms are around inclusion, challenging systems of power and inequality and attending to local, diverse perspectives, course materials should reflect this. Students need to see themselves reflected in the course materials to connect with it (Hurtado et al., 2012). Given that BIPOC and other marginalized groups author a great deal of critical environmentalisms-themed work in a variety of modes (e.g., podcasts, images, social media and film/video in addition to a wide-variety of written works), finding relevant material that both meets the goals of the class and students' interests is not hard. And incorporating a variety of modes of expression and communication is important to challenge the elitism of knowledge production and to incorporate a wider variety of contributors, as work that is more accessible and public also tends to be produced by more diverse authors (Portland Community College Library, 2014). Yet I would argue that it is critically important to incorporate this principle into course and curriculum design across all disciplines, especially in fields that are predominantly white. It can be enormously impactful: it provides a well-rounded curriculum, fosters greater inclusivity into our courses, overcomes some of the historical exclusions of different types of knowledges within academia and brings attention to the impacts of identity on knowledge production (Mott & Cockayne, 2017). Students see their backgrounds or identities and different forms of communication and styles of expression represented in the field. If disciplines are serious about diversifying their fields, they must begin with a diverse curriculum, which includes a broad array of content/perspectives, modes of delivery and authors.

Use Writing as a Tool to Facilitate Engagement, Communication, Comprehension and Analysis Most writing instructors are well-aware of and regularly integrate writing into their courses, not simply because their courses are focused on writing, but because writing helps to facilitate student engagement with each other and the course materials. Like the active listening exercise discussed above, writing slows the learning and communication process, so that students can process and reflect upon ideas, grapple with complexity or clarify what they don't understand. Writing exercises can also defuse a tense or escalating situation, and help students better understand multiple perspectives. Often, before we begin a class discussion, or in cases when a class discussion becomes heated, I ask students to do a one- or two-minute free-write to jot down what they are thinking and feeling or how they are responding to the course materials or to a specific prompt. Their free-writes reveal the complexity of the issues at stake. Once they've been given a moment to reflect, they can contribute to class discussion with more clarity and availability to hear others' perspectives.

Even in my non-writing intensive courses, I often assign both low- and high-stakes writing assignments. Low-stakes writing assignments include reading summaries and/or responses, journal articles, reflection papers, stream-of-consciousness free-writes, the minute paper and exit tickets. These activities can center a class, provide a starting point for class discussion, and allow students to collect their thoughts before entering into a faster paced discussion. Higher stakes writing assignments include formal papers that go through multiple drafts and stages of revision. For these activities, students respond to each other's drafts, learning from the process of reading others' work, learning about their thoughts and ideas and receiving feedback on their writing and ideas. This in turn builds rapport and support.

Give Students Some Say in the Design of the Course, and Regularly Solicit for (and When Possible, Integrate) Their Feedback I tell students that as the instructor, I am responsible for their courses, but all participants are responsible for the course's success. I provide a general curricular and structural framework for the course, but both are flexible and can be adjusted and tailored to student interests and needs within reason. This enables us to adapt to specific (and potentially trauma-inducing) contexts they, as a class, may be facing, like the multiple labor strikes, housing shortages or local wildfires UCSC students have faced recently, and gives students space to talk about and process their experiences and connect them to the course materials. This in turn reinforces course content by showing the ideas we are studying and methods we are practicing are grounded in our lived experiences. I invite students to provide regular, anonymized feedback about the course, and make suggestions about the design. Adapting the course to the needs and concerns of students every quarter requires more improvisation and on-demand labor, but it has important long-term pay-off: having a voice and a sense of agency and finding connections between others' ideas and their own experiences are skills that will serve them well outside of our class.

Recognize that Change Is Slow and Incremental, Including for Instructors Getting students to be more proactive and take charge of their education and learning can be difficult, especially challenging post-pandemic, as we re-learn how to relate to one another (Malesic, 2022), and in the age of digital technologies (Turkle, 2011). The point is not to socialize students into relating to one another in times before the pandemic or the advent of personal technologies. The point is that as we find new ways to relate to one another (and discover old ways), we prioritize and center relationality and a relational ontology in our interactions. Emphasizing the value of relationships renders the particular mode of interaction less important.

As the instructor, I do not require that students do anything they are uncomfortable with, but I gently challenge them to take risks and I create classroom conditions that support that. In best case scenarios, students understand and engage with both the pedagogy and content from the outset of the course. In other cases, one or more student(s) or even an entire class may resist this process despite my best efforts, and I've come to accept that too. Every class is unique, each with their own inexplicable alchemy and context that meets different course outcomes at different levels. If we take the idea of relationality seriously, we understand that the impacts of a particular class may not be realized until much later, once the class is over and even perhaps years later after a student has matured. As instructors, we have to be comfortable using this pedagogy and not expect immediate change, trusting the course's larger aims may be experienced later on down the road.

Like critical sustainability studies, relational pedagogies value flexibility, improvisation, and adapting to the needs of the moment; likewise, these suggestions are meant to be adapted to a particular set of circumstances—they are not a set of hard and fast rules. Small classes and/or seminar-style formats with a critical focus lend themselves most easily to designing around a relational pedagogy, yet relational principles can be integrated in a wide-range, if not all, course sizes and disciplinary subjects. These principles can inform curriculum design, instructional delivery, classroom management and orient them towards enhancing relationality and student engagement (Ambrose et al., 2010; Bain, 2004). And given that healthy relationships are shown to positively impact student learning and satisfaction (Felton & Lambert, 2020), I would encourage designing any course around these principles, even when a critical approach or anything related to human relationships is not the primary focus of the course.

Making any changes to any course can be labor-intensive and integrating more of a relational approach is no exception. This method also requires a degree of self-awareness—relational instructors need to be committed to an ongoing practice of self-reflection to ensure that they are coming to the class in the spirit of genuine collaboration and openness. But I would argue that putting in the effort to develop a relational pedagogy is worth it—teaching is far richer and more rewarding for students and instructors alike, as research demonstrates that fostering high quality relationships can lead to higher levels of life satisfaction (Vaillant, 2002). Relational pedagogies can model and teach students how to forge and sustain high quality relationships inside the classroom, which will ideally translate to relationships outside the classroom. It can bring back the joy and fulfillment in teaching and learning that many argue was lost during and in the aftermath of the pandemic.

If we focus on building more positive, resilient and satisfying relationships, not just in our courses, but in our institutions overall, we offset the stultifying bureaucracies and dehumanizing politics of these organizations. They provide alternatives to hyper-competitive and individualistic capitalist, race-to-the-bottom mentalities that are characteristic of neoliberalized approaches to the environment and higher education. Instead of focusing on institutional demands of content mastery, course outcomes or student achievement, relational pedagogies can reinvigorate teaching and bring back humanistic elements that are a key part of meaningful, sustained education. The more we can integrate these principles and practices into our classes, especially courses in critical environmentalism, the more students will walk away not just with content knowledge, but real-world relational skills and intrinsic motivation to work towards social justice outcomes that will serve them well in the long-term.

Conclusion

Integrating relational pedagogies when teaching critical sustainabilities and environmental courses has the potential to transform teaching and learning in these fields, such that students are empowered to work towards substantive, meaningful change and avoid reproducing Cartesian, Western hierarchies and power relations. Centering a relational ontology helps us reimagine human/non-human relations within a horizontal, networked framework, while drawing from pedagogical practices that foster and teach relational skills that give learners specific tools to operate within that framework. By centering a "relationships first" model in teaching, we help students create alternative models for being and acting in the world in accordance with their deepest convictions.

Perhaps most importantly, I want to suggest that a relational approach has profound implications not just for our students and classroom pedagogy, but for transforming the future of environmental disciplines themselves. The move towards equity, diversity and inclusion—re-centering historically underrepresented voices—is some of the most important work that can be done in this field. By broadening the scope of who is included, we increase the richness of our community and knowledge base. By listening to, learning from and incorporating the voices of historically underrepresented groups, all aspects of environmental fields, from methodological and research practices, to philosophical questions of epistemology and ontology, to pedagogy and beyond, have the possibility for radical transformation, strengthening them to become more flexible, adaptable and inclusive in the global age.

References

Agne, R. R., & Muller, H. L. (2019). Discourse strategies that co-construct relational identities in STEM peer tutoring. *Communication Education, 68*(3), 265–286.

Ambrose, S. A., Bridges, M. W., DiPietro, M., Lovett, M. C., & Norman, M. K. (2010). *How learning works: Seven research-based principles for smart teaching*. Wiley.

Aspelin, J. (2014). Beyond individualised teaching: A relational construction of pedagogical attitude. *Education Inquiry, 5*(2), 23926. https://doi.org/10.3402/edui.v5.23926

Aspelin, J. (2020). Teaching as a way of bonding: A contribution to the relational theory of teaching. *Educational Philosophy and Theory, 53*(6), 588–596. https://doi.org/10.1080/00131857.2020.1798758

Bain, K. (2004). *What the best college teachers do*. Harvard University Press.

Baker, J. A., Grant, S., & Morlock, L. (2008). The teacher-student relationship as a developmental context for children with internalizing or externalizing behavior problems. *School Psychology Quarterly, 23*(1), 3. https://doi.org/10.1037/1045-3830.23.1.3

Battle, C. P. (2022, April 4). *#WeChooseNow: Climate justice from the southern frontlines* [Public lecture]. Institute for Social Transformation, UCSC.

Biesta, G. (2004). "mind the gap!" communication and the educational relation. *Counterpoints, 259*, 11–22.

Bingham, C. W., & Sidorkin, A. M. (2004). *No education without relation*. Sacramento State.

Brune, M. (2020). *Pulling down our monuments*. https://www.sierraclub.org/michael-brune/2020/07/john-muir-early-history-sierra-club

Buber, M. (2002). *Between man and man*. Routledge.

Caine, J., Gilroy, J., Greaves, M., & Madriaga, M. (2022). Principles of success: Facilitating sustainable transformation through a progressive relational pedagogy. *Social Policy and Society, 21*(1), 106–122. https://doi.org/10.1017/S1474746421000658

Cammarota, J. (2006). A critically compassionate pedagogy for Latino youth. *Latino Studies, 4*(3), 305–312.

Cammarota, J., & Romero, A. (2006). A critically compassionate intellectualism for Latina/o students: Raising voices above the silencing in our schools. *Multicultural Education, 14*(2), 16–23.

Chika-James, T. A. (2020). Facilitating service-learning through competencies associated with relational pedagogy: A personal reflection. *Higher Education Pedagogies, 5*(1), 267–293. https://doi.org/10.1080/23752696.2020.1820886

Claborn, J. (2014). WEB Du bois at the grand canyon. *The Oxford handbook of ecocriticism*, 118–131.

Jones, J., & Crownover, A. (2016). *Constructing a relational pedagogy: Teacher educators' role in reconceptualizing how bullying is addressed in schools*. Teachers College Record.

Darby, L. (2005). Science students' perceptions of engaging pedagogy. *Research in Science Education, 35*(4), 425–445.

Edwards, J. B., & Richards, A. (2002). Relational teaching: A view of relational teaching in social work education. *Journal of Teaching in Social Work, 22*(1–2), 33–48. https://doi.org/10.1300/J067v22n01_04

Fears, D., & Mufson, S. (2020, July 22). Liberal, progressive—And racist? The Sierra Club faces its white supremacist history. *The Washington Post*. https://www.washingtonpost.com/climate-environment/2020/07/22/liberal-progressive-racist-sierra-club-faces-its-white-supremacist-history/

Felton, P., & Lambert, L. (2020). *Relationship rich education: How human connections drive success in college*. John Hopkins University Press.

Ferreira, F. (2017). Critical sustainability studies: A holistic and visionary conception of socio-ecological conscientization. *Journal of Sustainability Education, 13*, 1–22.

Field Station Five. (2020). *Black Feminist Interventions in Environmentalism with Chelsea Frazier* [Podcast]. https://www.anthropocene-curriculum.org/contribution/broadcasting-live-from-field-station-5-episode-2

Foster, J. B. (2000). *Marx's ecology: Materialism and nature*. New York University Press.

Freire, P. (2007). *Pedagogy of the oppressed: 30th anniversary edition*. The Continuum International Publishing Group.

Graff, G. (1994). The pedagogical turn. *The Journal of the Midwest Modern Language Association, 27*(1), 65–69.

Greenberg, M. (2013). What on earth is sustainable? Toward critical sustainability studies. *Boom: A Journal of California, 3*(4), 54–66. https://doi.org/10.1525/boom.2013.3.4.54

Gross, N., & Lo, C. (2018). Relational teaching and learning after loss: Evidence from Black adolescent male students and their teachers. *School Psychology Quarterly, 33*(3), 381–389. https://doi.org/10.1037/spq0000285

Hickey, A., & Riddle, S. (2021). Relational pedagogy and the role of informality in renegotiating learning and teaching encounters. *Pedagogy, Culture & Society*, 1–13. https://doi.org/10.1080/14681366.2021.1875261

Hurtado, S., Alvarez, C. L., Guillermo-Wann, C., Cuellar, M., & Arellano, L. (2012). A model for diverse learning environments: The scholarship on creating and assessing conditions for student success. *Higher Education: Handbook of Theory and Research, 27*, 41–122.

Kimmerer, R. (2013). *Braiding sweetgrass: Indigenous wisdom, scientific knowledge and the teachings of plants*. Milkweed editions.

Klem, A. M., & Connell, J. P. (2004). Relationships matter: Linking teacher support to student engagement and achievement. *Journal of School Health, 74*, 262–273.

Latour, B. (2007). *Reassembling the social: An introduction to actor-network-theory*. Oxford University Press.

Longino, H. E. (2017). *Feminist epistemology. The Blackwell guide to epistemology* (pp. 325–353).

Lu, F., Rosser, R. H., Renteria, A., Kim, N., Erickson, E., Sher, A., & O'Connor, L. (2018). Inclusive sustainability: Environmental justice in higher education. In W. Leal Filho, R. W. Marans, & J. Callewaert (Eds.), *Handbook of sustainability and social science research* (pp. 63–81). Springer.

Malesic, J. (2022, May 13). My College Students Are Not OK. *The New York Times*. https://www.nytimes.com/2022/05/13/opinion/college-university-remote-pandemic.html

Mott, C., & Cockayne, D. (2017) Citation matters: mobilizing the politics of citation toward a practice of 'conscientious engagement.' Gender, Place & Culture, 24:7, 954–973. https://doi.org/10.1080/0966369X.2017.1339022.

Murai, R. (2022, Feb 18). My family lost our farm during Japanese incarceration. I went searching for what remains. *Mother Jones*. https://www.motherjones.com/environment/2022/02/my-family-lost-our-farm-during-japanese-incarceration-i-went-searching-for-what-remains/

Noddings, N. (1992). In defense of caring. *The Journal of Clinical Ethics, 3*(1), 15–18.

Noddings, N. (2005). Identifying and responding to needs in education. *Cambridge Journal of Education, 35*(2), 147–159.

Papenfuss, J. T. (2019). *Inside-out pedagogies: Transformative innovations for environmental and sustainability education* (Doctoral dissertation, Arizona State University).

Pellow, D. N. (2016). Toward a critical environmental justice studies: Black lives matter as an environmental justice challenge. *Du Bois Review: Social Science Research on Race, 13*(2), 221–236. https://doi.org/10.1017/S1742058X1600014X

Peet, R., & Watts, M. (2004). *Liberation ecologies: Environment, development and social movements*. Routledge.

Pepper, D., Webster, F., & Revill, G. (Eds.). (2003). *Environmentalism: Critical concepts* (Vol. 1-4). Taylor & Francis.

Portland Community College Library. (2014). *Know your sources* [Infographic]. Retrieved January 28, 2023 from https://www.pcc.edu/library/scripts/know-your-sources/index.html

Prosser, M., & Trigwell, K. (1999). Relational perspectives on higher education teaching and learning in the sciences. *Studies in Science Education, 33*(1). https://doi.org/10.1080/03057269908560135

Purdy, J. (2015). Environmentalism's racist history. *The New Yorker, 13*.

Reeves, J., & Le Mare, L. (2017). Supporting teachers in relational pedagogy and social emotional education: A qualitative exploration. *International Journal of Emotional Education, 9*(1), 85–98.

Reichert, M., & Hawley, R. (2013). Relationships play primary role in boys' learning. *Phi Delta Kappan, 94*(8), 49–53.

Sneddon, C. S. (2000). 'Sustainability' in ecological economics, ecology and livelihoods: A review. *Progress in Human Geography, 24*(4), 521–549.

Trauth-Nare, A., Buck, G. A., & Beeman-Cadwallader, N. (2016). Promoting student agency in science inquiry: A self-study of relational pedagogical practices in science teacher education. In G. A. Buck & V. L. Akerson (Eds.), *Enhancing professional knowledge of pre-service science teacher education by self-study research* (pp. 43–67). Springer.

Turkle, S. (2011). *Alone together : Why we expect more from technology and less from each other*. Basic Books.

Vaillant, G. E. (2002). *Aging well: Surprising guideposts to a happier life from the landmark study of adult development*. Little.

Warren, C. A. (2014). Towards a pedagogy for the application of empathy in culturally diverse classrooms. *The Urban Review, 46*(3), 395–419.

Warren, C. A., & Lessner, S. (2014). "Who has family business?" Exploring the role of empathy in student-teacher interactions. *Penn GSE Perspectives on Urban Education, 11*(2), 122–131.

Wentzel, K. R. (2009). Students' relationships with teachers as motivational contexts. In K. R. Wentzel & A. Wigfield (Eds.), *Handbook of motivation at school* (pp. 315–336). Routledge.

Whatmore, S. (2002). *Hybrid geographies: Natures cultures spaces*. Sage.

Whyte, K. (2017). Indigenous climate change studies: Indigenizing futures, decolonizing the Anthropocene. *English Language Notes, 55*(1), 153–162.

Wildman, W. J. (2010). *An introduction to relational ontology. The trinity and an entangled world: Relationality in physical science and theology* (pp. 55–73). Eerdmans.

Emily Murai is a Lecturer at College Nine and the Environmental Studies department at the University of California, Santa Cruz. She teaches courses on critical reading, writing and thinking, writing-across-the curriculum/writing-in-the-disciplines (WAC/WID), and in the environmental humanities. She holds a Ph.D. in Human Geography from the University of Minnesota, an M.A. in Sociology from the New School for Social Research and a B.A. in American Studies and Psychology from UCSC.

Part II

UCSC Students' Sustainability Perceptions, Understandings, and Values

4 Student Understandings of Sustainability

Flora Lu, Tashina J. Vavuris, Randy Uang, Cassandra Wood, and Anna Sher

Abstract

This chapter explores 'sustainability' as a socially produced, contested, and contextually dependent concept, with a focus on unpacking its simultaneously evident and evasive meanings. The notion of sustainability is applied to a range of activities and approaches, as a means of emphasizing moral commitment to environmental values, future orientation, and a higher good. Members of university communities have often taken on the mantle of sustainability work and education, while taking for granted the terms, targets and theorizations behind these efforts. Drawing on campus-wide surveys of undergraduates conducted in 2019 and 2022, this chapter first examines how undergraduates conceptualize sustainability, and then how different definitions rank depending on students' characteristics. We find that a plurality of definitions are held, with interesting discrepancies by race, gender, educational status, and major. We argue that critical sustainability education is fostered when students interrogate the meanings they and their peers hold.

F. Lu (✉) · T. J. Vavuris · R. Uang · A. Sher
University of California, Santa Cruz, CA, USA
e-mail: floralu@ucsc.edu

C. Wood
University of California, Los Angeles, CA, USA

Keywords

Sustainability · Plurality of understandings · Race · Gender · First-generation status · Major discipline

Introduction

Sustainability is aspirational, normative, and visionary, a commitment that everyone purports to support. However, it is not fixed or monolithic, but rather a socially constructed concept with multiple definitions and understandings (Emanuel & Adams, 2011; Rose & Cachelin, 2018). As Scoones (2007: 589) reminds us, sustainability is a buzzword with ambivalent and contested meanings, and striving towards it necessitates efforts to create "epistemic communities of shared understanding and common commitment." However, the debate has been dominated by economic and environmental facets of sustainability (Colantonio, 2009), from the concept's beginnings in forestry in 1712, to the definition of sustainable development in the iconic Brundtland report, *Our Common Future*, to the theorizations of mathematical ecologists and neo-classical economists, to the idea of a "triple bottom line" in business vernacular. Multiple versions of sustainability—"broad and narrow, strong and weak, big S and small s" (Scoones, 2007: 590)—compete with each other in confusing discourse. Sze

F. Lu, E. Murai (eds.), *Critical Campus Sustainabilities*, Sustainable Development Goals Series,
https://doi.org/10.1007/978-3-031-30929-8_4

et al. (2018: 3) note that the ambiguity of the term sustainability "has contributed to its large-scale acceptance" as a framework for action, but that efforts toward sustainability must contextualize, situate and reject such vagueness in order to be successful. The point is not to settle on one general, universal definition of sustainability, but instead grapple with how various interpretations reflect power dynamics, agendas, and world views that must be recognized and reconciled. Leach et al. (2010: 5) argue that society must push back on the "pervasive tendency...for powerful actors and institutions to 'close down' around particular framings, committing to particular pathways that emphasize maintaining stability and control...This can in turn obscure or deny the reality of alternatives." Dillon and Sze (2018: 263) allude to notions of sustainability that entangle "race, class, and other relations of power in the task of building better, more liveable worlds." As educators and learners, these complex, contested and concurrent framings of sustainability constitute a useful and necessary space for critically interrogating these interpretations and the interests, goals, trade-offs, and visions for the futures they represent.

Scholarship about undergraduate college students' conceptualizations of sustainability highlight the need for further study and improved pedagogical efforts. In an online survey of 1889 undergraduates at the University of Plymouth in southwest England, Kagawa (2007) found that over 90% of them held a positive perception of sustainability or identified themselves as passionate advocates for it; both students who were familiar and unfamiliar with the concept of sustainability deemed it "a good thing." Kagawa noted that such general support for sustainability occurred without an understanding of its contested and multifaceted meanings. Furthermore, the term was also equated with environmental concerns and only marginally with social, political, economic and cultural dimensions. Such knowledge or concept deficits were found in other studies about the sustainability perceptions of university students (Azapagic et al., 2005; Summers et al., 2004). While University of Plymouth respondents tended to agree with the urgency of changing ways of living to offset the danger of climate change, the lifestyle changes they deemed acceptable tended towards consumerist, individual choices such as recycling, buying green or conserving water ("light green" actions), rather than structural, institutional and political changes ("dark green" responses).

Other studies about students' sustainability perceptions compared institutions. For example, Emanuel and Adams (2011) examined understandings of sustainability among a random sample of 406 undergraduates at a university in Alabama and one in Hawaii. The authors operationalized the concept of sustainability in terms of facilities and infrastructure, waste, energy use, and campus commitment. Emanuel and Adams found that about 38% of respondents in Alabama and 32% of respondents in Hawaii said that they do not know much about sustainability, although the latter scored much higher on knowledge questions and statements expressing a willingness to work towards campus sustainability (what the authors characterized as a "commitment gap"). Bajracharya and Maskey (2016) implemented a survey exploring beliefs, knowledge, perceptions and behaviors around sustainability with 373 students at a university in West Virginia and one in Central New York. While institutional differences were slight, the authors did find marked differences in sustainability perceptions by gender and major.

Within academic literature, student perceptions of sustainability are an under-researched area (Emanuel & Adams, 2011). In this chapter, we approach the topic differently from the aforementioned studies by focusing on various understandings of sustainability that move beyond the dominant discourse. We present five definitions of sustainability, along a gamut of ecological, economic, infrastructural, social, and cultural dimensions, with an orientation on structural, "dark green" responses rather than individual behavioral or consumer choices. They are:

Using Natural Resources Wisely This definition refers to a core idea about sustainability, as articulated in the Merriam Webster Collegiate Dictionary: "relating to, or being a method of harvesting or using a resource so that the resource is not depleted or permanently damaged." The association of sustainability with renewable harvest beckons to the natural science origins of the term, specifically the emergence of sustainable yield forestry in the early twentieth century and names such as Gifford Pinchot, the first director of the US Forest Service (Whitehead, 2014: 258). There is a utilitarian managerialism reflected in this definition, largely devoid of politics, equity concerns, complexity and dynamism. This definition also simplifies the natural environment as 'resources'.

Building Green Infrastructure This definition was influenced by the University of California, Santa Cruz (UCSC) Sustainability Office, which incorporates a multi-pronged approach towards green infrastructure, from LEED certification for buildings, solar panels and increasing electric vehicle charging stations, to increasing the use of non-potable water on campus and prioritizing shared transit and human-powered transportation. The Environmental Protection Agency envisions green infrastructure as an integral part of sustainable communities. For example, a network of parks in Los Angeles Country constructed a bioswale filled with drought-tolerant native plants to infiltrate stormwater, added picnic areas and benches, and installed interpretive signage in the Tongva Indian language (EPA, 2014).

Challenging Over-Exploitive Economic Structures This aspect of sustainability addresses the need to recognize the ways in which capitalism, as a form of economic and social organization, renders imperative capital accumulation, leading to privatization, commodification and the "enclosure of biophysical and other resources in process of accumulation by dispossession" (Feola, 2020: 242; Harvey, 2006). Growing inequality, the externalization of environmental degradation, and the fallacy of endless growth run counter to the goal of a more sustainable society, yet research on sustainability transitions has yet to more deeply engage critiques of capitalism (Feola, 2020). Greater attention is being paid to social movements in the Global South that offer an alternative, non-capitalist vision of living well (e.g., *buen vivir* and *sumak kawsay* in Latin America, Lu et al., 2017; Escobar, 2010), a "radical questioning of the core assumption of growth and economism, [and] a vision of alternative worlds based on ecological integrity and social justice" (Escobar, 2015: 451).

Promoting Equity In his articulation of critical environmental justice studies, Pellow (2018) argues that environmental equity not only entails procedural and distributional justice, but deeper considerations of the ways that gender, sexuality, and other categories of difference shape struggles over sustainability. It is not enough to merely to document how food insecurity, lack of greenspace, contamination, etc. are disproportionately experienced by low income, people of color, who also often lack a seat at the table of decision making on environmental policy—advocates for environmental justice must emphasize the systems that give rise to and perpetuate such inequities. An equity lens understands that "racism, heteropatriarchy, classism, nativism, ableism, ageism, speciesism...and other forms of inequality...are mutually reinforcing...[and] produce and maintain systems of individual and collective power, privilege and subordination" (Pellow, 2018: 19). Pellow stresses the need for analyses that are intersectional and multiscalar. He calls on the need to confront the entrenchment of power and inequality in the state and society, and promote an ideology of racial and socio-ecological indispensability.

Centering Cultural Knowledge This definition includes cultural stewardship practices, especially those of Indigenous people, that are increasingly recognized as pathways for effective environmental conservation and sustainability (Artelle et al., 2019; Ban et al., 2018; Berkes, 2018). Sustainability must recognize the interdependence between humans and ecosystems; the natural world both shapes and is shaped by humans, who have developed systems of meaning, knowledge, governance to coexist with and care for the environment (Balée, 2013; Berkes, 2018; Denevan, 1992; Fa et al., 2020; Norman, 2017; Whyte & Cuomo, 2016). This definition points to concurrent processes in which diversity flourishes in socio-ecological systems, such as the congruences between biodiversity, cultural diversity, and linguistic diversity (Maffi, 2005).

In 2019 and 2022, the People of Color Sustainability Collective (PoCSC), led by authors Lu and Vavuris, and the Institutional Research, Assessment and Policy Studies (IRAPS), led by authors Sher and Uang, implemented campus-wide, undergraduate surveys to better understand students' perceptions of, behaviors toward, and experience of sustainability. These surveys were part of a larger effort to develop more effective co-curricular programming towards greater environmental inclusion and equity. As part of the survey instrument, students were asked to rank these five definitions of sustainability and explicate their reasons for doing so. In this chapter, we present key findings from these two surveys, which were taken by over 4800 students, a sample size significantly higher than many other studies of this kind. The results demonstrate a plurality of understandings of sustainability that, while reflecting dominant understandings of the term, reveal how race, gender, first generation status, and academic discipline offer different priorities and interpretations.

Methods

This chapter is based on surveys of UCSC undergraduates conducted in spring 2019 and 2022. All enrolled students (16,983 undergraduates in 2019 and 17,064 in 2022) were invited to take an online survey in which they were asked to share their perceptions of and participation in the environmental sustainability movement on campus. The survey response rates were 15% (n = 2546) in 2019 and 13% (n = 2273) in 2022. The authors worked closely together on survey design, administration, and data analysis. This research was approved by the Institutional Review Board. Survey respondents were representative of the student population in terms of race and ethnicity, first generation status, transfer status, and class level. Weighted data was used for our analysis to ensure proper representation of the student population in terms of gender and ethnicity. In addition to the survey data, the analyses incorporated institutional data on students' demographic background (gender, race/ethnicity, first generation status) and academic path (major, college, class level, and transfer status).

To better understand how survey respondents understand the concept of sustainability, we asked them to rank five definitions of the term in descending order of importance, with a rank of 1 being the most important, and 5 as the least important. The five definitions spanned a range of different conceptualizations of sustainability, and the list of definitions was presented randomly to each student, to avoid item order effects. Below we give the definitions and the short-hand term we will use to report our findings.

- Sustainability is building green facilities and infrastructure. (BUILDING)
- Sustainability is using natural resources wisely. (USING)
- Sustainability is challenging societal economic structures that overexploit the environment. (CHALLENGING)
- Sustainability is promoting culturally diverse knowledges and relationships with the environment. (PROMOTING)
- Sustainability is working toward social, economic, and environmental equity. (WORKING)

After asking participants for their rankings, we then asked them a follow up, open-ended question of why they ranked the sustainability definitions as they did. Students' responses offer some insights into how the various choices were interpreted and justified. Comments around the BUILDING definition are rather straightforward, such as the comment, "We need to build more infrastructure that won't add harmful things to the environment." For the USING definition, one student wrote, "We need to find a way to use the least resources possible to get the most benefit for us while doing the least amount of damage to the planet." This response highlights a cost/benefit calculation, a maximization of utility for people while minimizing harm to the environment. For the CHALLENGING definition, another student explained, "The environment's biggest threat is big business and politicians who wield a disproportionate amount of power, and inherently benefit from the destruction of the environment. Nothing large-scale can be accomplished unless those in power are pro-sustainability." Responses explaining why some students ranked PROMOTING as an important understanding of sustainability included, "Learning from people's practices to steward the earth. The world is diverse and we need to know how to take care of it with diversity." Finally, equity considerations alluded to in the WORKING definition was understood by one respondent as, "Equity is essential for sustainability, because efforts aren't sustainable unless they are sustainable for everyone in all demographics...not just wealthy white people who can afford to be environmentally conscious." Multiple valences of the term sustainability show up here, i.e., environmental care and stewardship, something that can be maintained over time, and notions of self-efficacy and opportunity/accessibility.

In 2019, a total of 2014 students answered the questions around definitions of sustainability: this breaks down racially as 1225 people of color (PoC) and 624 white, non-Hispanics (WNH). We also compared the rankings of male (n = 1028) versus female (n = 935) respondents; first generation college students (n = 708) versus non-first-generation students (n = 1308); and Environmental Studies and Sciences majors (n = 180) versus students majoring in other fields (n = 1834). In 2022, a total of 1688 students answered the questions around definitions of sustainability: 1016 PoC and 557 WNH (the remaining were international students, and were not included in this analysis). Again, we compared the rankings of male (n = 843) versus female (n = 774) versus nonbinary/genderqueer (n = 56) identified respondents; first generation college students (n = 545) versus non-first-generation students (n = 1143); and Environmental Studies and Sciences majors (n = 151) versus students majoring in other fields (n = 1537). In the analysis, we utilized descriptive statistics to characterize what proportion of these groups ranked various definitions as their top two conceptualizations of sustainability, and then ran statistical tests (chi-square test for dichotomous categorical variables) to determine if the differences were statistically significant.

Findings

Overall Campus Rankings

Overall, in 2019, the UCSC undergraduates as a whole ranked USING as the top definition of sustainability (Fig. 4.1)—this option had a mean ranking of 2.65 and 50% of respondents ranked it as either 1 or 2. In second place, the CHALLENGING definition had a mean ranking of 2.75 and 46% of respondents ranked it 1 or 2. Equity considerations captured in the WORKING definition were deemed third most important, with a mean rank of 2.98 and 41% of respondents ranking it 1 or 2. There is a gap between the top three most important definitions and the remaining two. BUILDING and PROMOTING have a mean rank of 3.3 and 3.33, with only 33% and 31% of respondents, respectively, ranking them 1 or 2. While the determination of the least most favored definition is close, the option that refers to epistemological breadth and cultural salience in addressing sustainability falls into fifth place.

In 2022 (Fig. 4.2), we found that the rank order of the first two and the last two definitions changed. Instead of USING, the top ranked definition was CHALLENGING. Whereas PROMOTING was the fifth ranked definition in 2019, in 2022, BUILDING was the lowest, with PROMOTING rising to fourth place. Note that while the ordering of the definitions changed, the percentages and mean rankings did not change much between 2022 from 2019.

In the remainder of this chapter, we compare how the rankings of these five definitions of sustainability differ between students based on race,

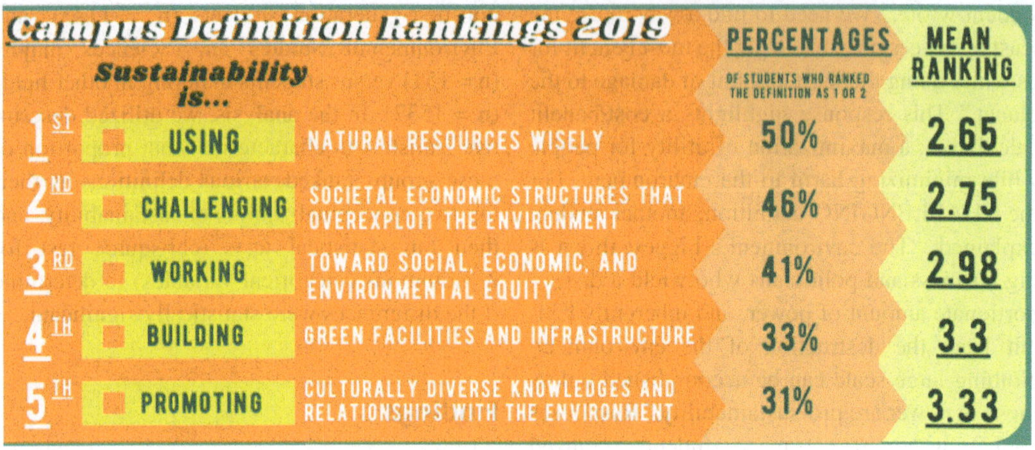

Fig. 4.1 Overall campus-wide rankings of the five sustainability definitions in 2019

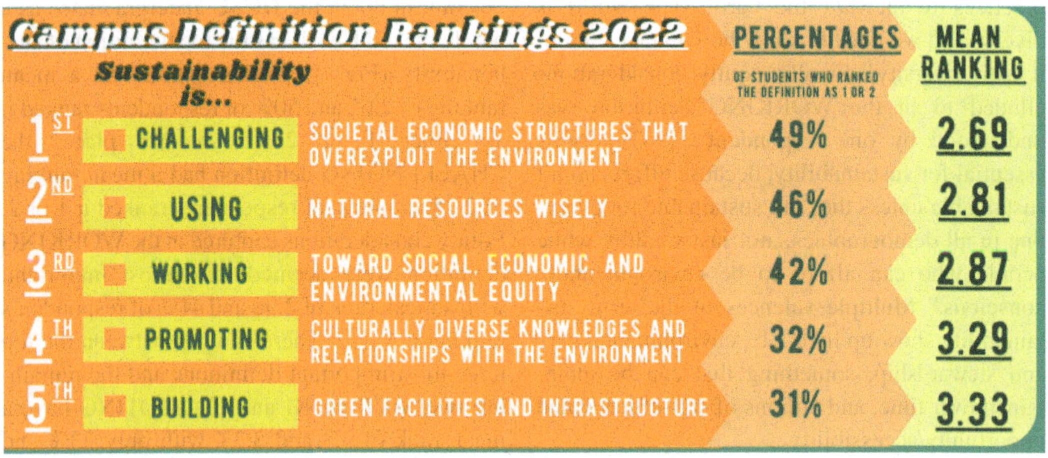

Fig. 4.2 Overall campus-wide rankings of the five sustainability definitions in 2022

gender, first-generation status, and major of study. Furthermore, we explore whether and how these results changed between 2019 and 2022.

Rankings Across Racial Categories

We compared the sustainability rankings of people of color (PoC) versus white, non-Hispanic (WNH) students (Figs. 4.3 and 4.4) in both 2019 and 2022.

In 2019, we found that the definitions were ranked in the same order (Fig. 4.3), but that the percentages sometimes differed between PoC and WNH. Two statistically significant differences ($p < 0.05$) were found for the first ranked and fifth ranked definitions. WNH students ranked USING higher than PoC, and PoC ranked PROMOTING higher than WNH, but not at high enough proportions to move it out of last place among students overall. It appears that economic definitions of sustainability (e.g., judicious use of resources and challenging over-exploitive economic systems) resonated more with WNH students than PoC, whereas definitions alluding to equity and cultural diversity were more highly ranked for PoC than WNH.

In 2022 (Fig. 4.4), the sustainability definition rankings between PoC and WNH were almost the same, with the difference that PROMOTING was

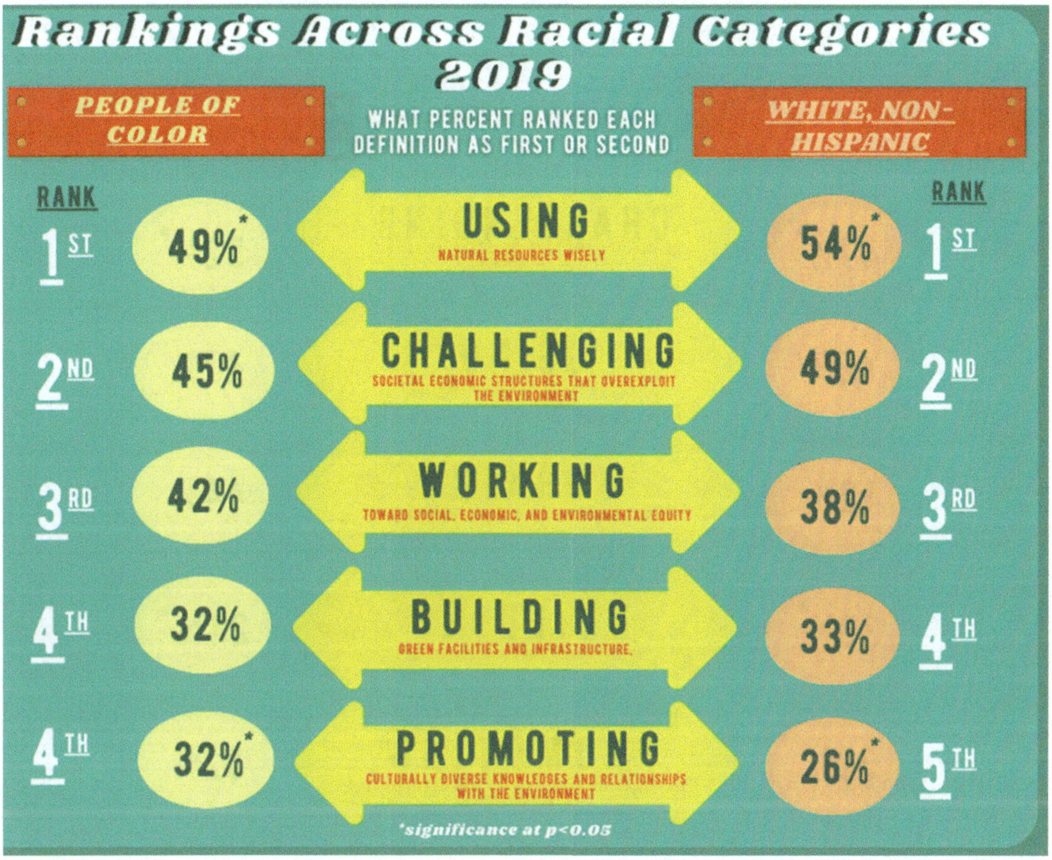

Fig. 4.3 Comparison of rankings of sustainability definitions between people of color and white, non-Hispanic students in 2019

the fifth ranked for PoC and BUILDING was the fifth ranked for WNH. For both PoC and WNH, CHALLENGING rose to the top as the most highly ranked definition, but more WNH students ranked it highly (52%) compared to PoC (47%), a statistically significant difference ($p < 0.05$). Another statistically significant result was that more PoC ranked BUILDING highly than did WNH ($p < 0.001$).

While the rankings for BUILDING and PROMOTING remained the same for PoC in 2019 and 2022, a major change emerged for WNH in regards to these two definitions, which essentially flip flopped in ranking between 2019 and 2022. For WNH respondents, PROMOTING went from 26% in 2019 to 32% in 2022; and BUILDING from 33% in 2019 to 24% in 2022. To summarize, the major racial differences we see in 2022 compared to 2019 is that WNH students moved away from USING as their top definition, towards CHALLENGING, and expressed more of an interest in PROMOTING diverse cultural knowledges and relationships with the environment, moving them to the same percentage as PoC.

In 2022, we conducted further analysis disaggregating the category of PoC (Table 4.1). Note that given their low numbers (<10), the results for Pacific Islanders are not shown. We found that the only statistically significant differences were between the categories of Hispanic/Latinx students and Asian American students, a reflection of their larger sample sizes. A higher percentage of Asian American students ranked highly the definitions of USING (51%) and BUILDING (38%), compared to Hispanic/Latinx students

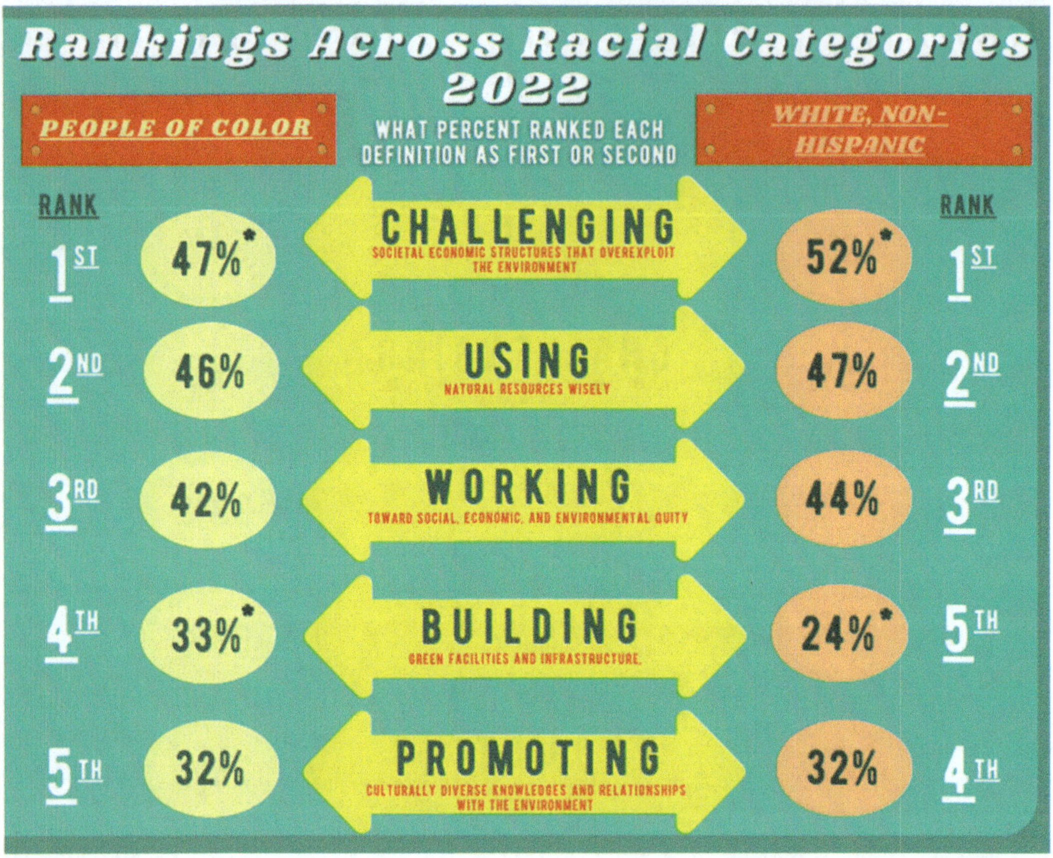

Fig. 4.4 Comparison of rankings of sustainability definitions between people of color and white, non-Hispanic students in 2022

(42% ranking USING and 28% ranking BUILDING highly). More Hispanic/Latinx students ranked the definition of CHALLENGING highly (53%), compared to Asian American students (42%). The percentage of white students ranking CHALLENGING highly (52%) was similar to that of Hispanic/Latinx students (53%).

Rankings Across Gender

In Academic Year 2018–2019, the undergraduate student population at UC Santa Cruz was 48.4% female and 50.7% male. Only two of the five sustainability definitions were statistically different between the two groups (Fig. 4.5). Female undergraduates place more importance than male students in understandings of sustainability CHALLENGING societal economic structures that overexploit the environment. Males place more value in BUILDING green infrastructure, whereas that definition of sustainability is the least important for females.

In Academic Year 2021–2022, the undergraduate student population at UC Santa Cruz was 45.9% female and 50.2% male and 3.9% non-binary/other. In reporting the results for the 2022 survey, we wanted to include the data for non-binary and genderqueer students. Because the numbers are relatively low compared to males and females, the differences in results of non-binary students were not statistically significant.

Figure 4.6 reports the gender comparison of sustainability definition rankings for 2022. The

Table 4.1 Disaggregation of sustainability definition rankings in 2022 for People of Color

	African American n = 81	American Indian/Alaskan Native n = 11	Asian American (AsA) n = 443	Hispanic/Latinx (H/L) n = 475	Significant differences
CHALLENGING	43%	38%	42%*	53%*	H/L > AsA
USING	45%	30%	51%*	42%*	AsA > H/L
WORKING	49%	38%	39%	44%	–
BUILDING	29%	37%	38%*	28%*	AsA > H/L
PROMOTING	33%	57%	30%	32%	–

* $p < 0.05$

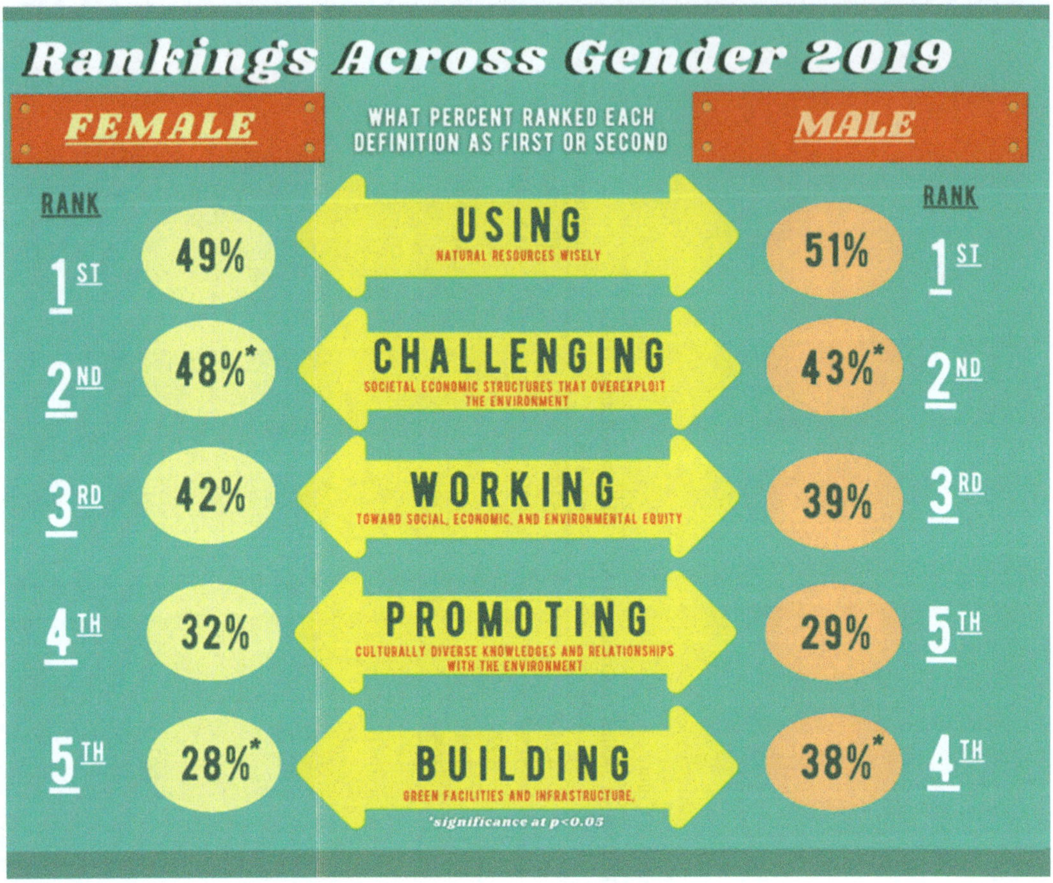

Fig. 4.5 Comparison of rankings of sustainability definitions by gender in 2019

rankings for females and non-binary students were the same, with CHALLENGING first, followed by WORKING, USING, PROMOTING and BUILDING. For males, however, USING was first, followed by CHALLENGING, WORKING, BUILDING and PROMOTING. A higher percentage of females than males prioritized WORKING ($p < 0.01$) and CHALLENGING ($p < 0.05$); a higher percentage of males ranked BUILDING ($p < 0.001$) and USING ($p < 0.01$) as first or second as their preferred definition of sustainability compared to females.

Comparing 2019 and 2022, we found that males and females diverged significantly over time. For males, the rank ordering of the definitions have remained consistent across the two surveys, but for females, there is a de-emphasis on USING, which moves from first place to third place. The statistically significant gender differences continued and expanded from 2019 to 2022: males still prioritize BUILDING more, but also USING too; females still prioritize CHALLENGING, but in 2022 diverged from males in terms of valuing WORKING more.

Rankings Across First-Generation Status

In Academic Year 2018–2019, 38.1% of the undergraduate student population at UC Santa Cruz was first-generation, i.e., neither parent(s)/guardian(s) received a four-year college degree; 59.3% were non-first-generation, or continuing generation. Wilbur and Roscigno (2016) drew upon a nationally representative longitudinal study to examine the challenges that first-generation students face. They found that first-

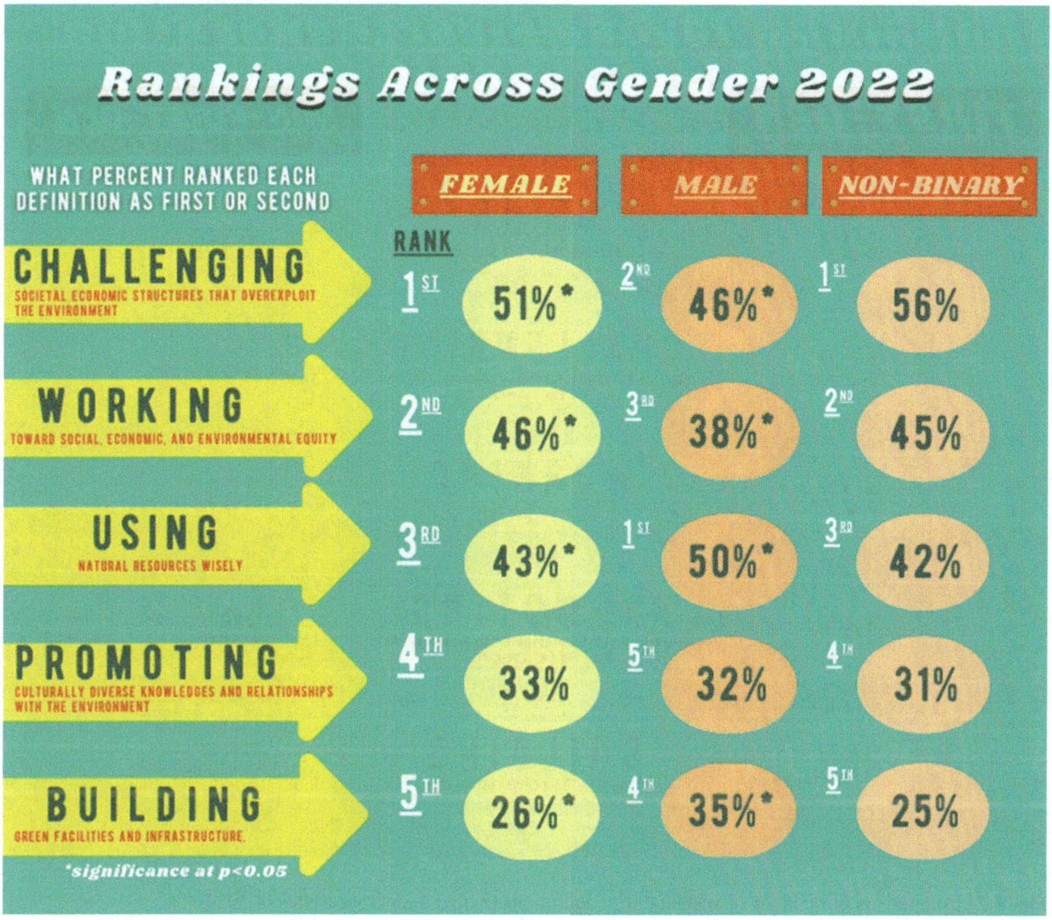

Fig. 4.6 Comparison of rankings of sustainability definitions by gender in 2022

generation students are 70% less likely than non-first-generation students to enroll in a four-year college and 60% less likely to graduate. First-generation students are more likely to work significantly more hours/week than their peers and to experience personal and family-related stressful life events. Socioeconomic status explains about a quarter of the unique disadvantages faced by first-generation students.

Figure 4.7 reports the results by first-generation status in the 2019 survey. We found that the top three definitions were statistically the same for first-gen and non-first-gen students, but first-generation students placed more importance on PROMOTING culturally diverse understandings and relationships with the environment than non-first-gen students, whereas non-first-gen students ranked BUILDING green infrastructure higher.

In Academic Year 2021–2022, 34.8% of undergraduates were first-generation and 62.6% were continuing generation. Figure 4.8 presents the comparison for first-generation vs. continuing generation students in 2022, finding only one statistically significant difference: that first-generation students prioritize WORKING toward social, economic and environmental equity more than continuing students ($p = 0.022$), ranking it second rather than third.

Comparing 2019 and 2022, we found a shift in the definitions that were different between first-generation vs. continuing students, from BUILDING and PROMOTING in 2019, to WORKING in 2022.

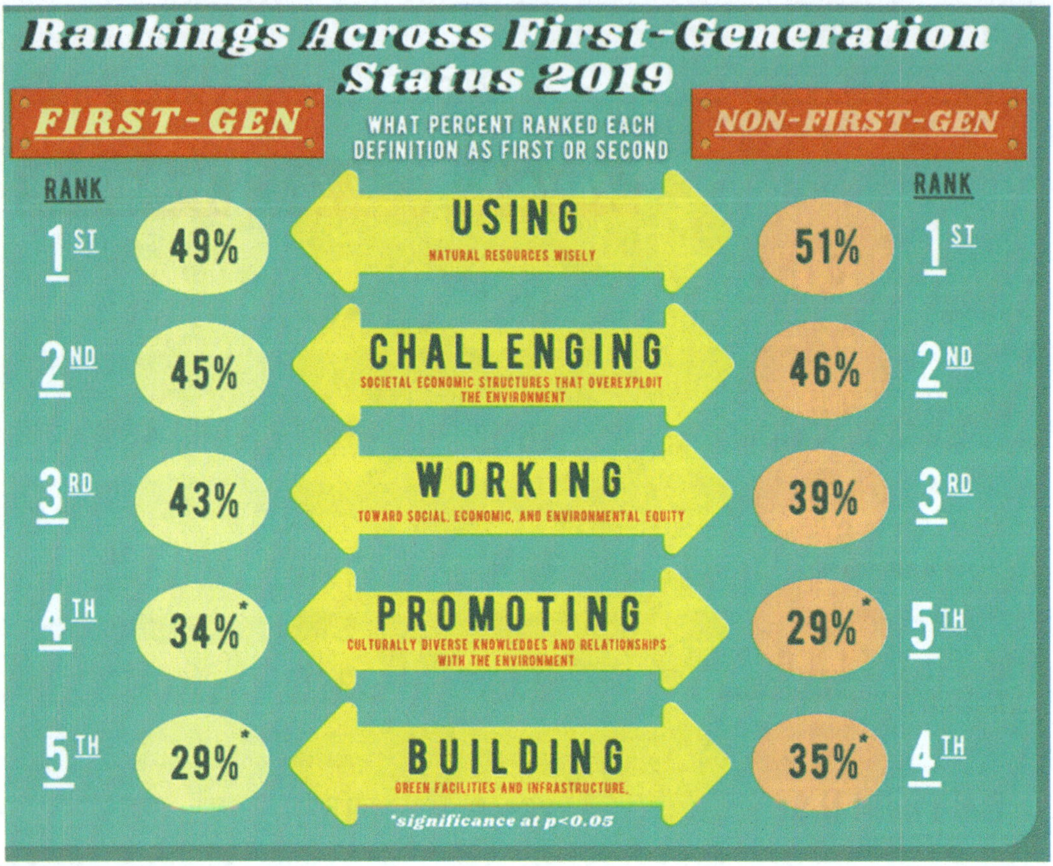

Fig. 4.7 Comparison of rankings of sustainability definitions by first-generation status in 2019

Rankings Across Disciplinary Major

Environmental Studies (ENVS) is an interdisciplinary department with approximately 500–600 proposed and declared majors. It offers a B.A. in Environmental Studies (with possible concentrations in Geographic Information Systems, Global Environmental Justice, and Conservation Science and Policy), a B.A. in Agroecology, and combined majors with Biology, Economics, and Earth Sciences. In this calculation, we include students majoring in Environmental Sciences (ENCI) as well, which is an interdepartmental, undergraduate-only degree program focused on Earth & Planetary Sciences and Ocean Sciences. We explored whether ENVS students differed in their prioritization of sustainability definitions compared to non-ENVS majors (Fig. 4.9). While only one of the comparisons between ENVS majors and non-ENVS majors was statistically significant—that ENVS students valued sustainability definitions that focused on equity considerations more—the patterns that emerged present the most striking contrast of the pairwise comparisons in 2019. Non-ENVS majors follow all other groups in ranking USING as the top definition, whereas ENVS majors ranked CHALLENGING and WORKING as tied in first place. This points to issues of structural overexploitation of the environment and the importance of equity, rather than resource use, as the predominant way that students who select ENVS view sustainability. Furthermore, it seems that ENVS students value cultural considerations over infrastructure, whereas the opposite pattern is the case for non-ENVS students.

In 2022 (Fig. 4.10), among ENVS majors, we see a three-way tie for first place for the

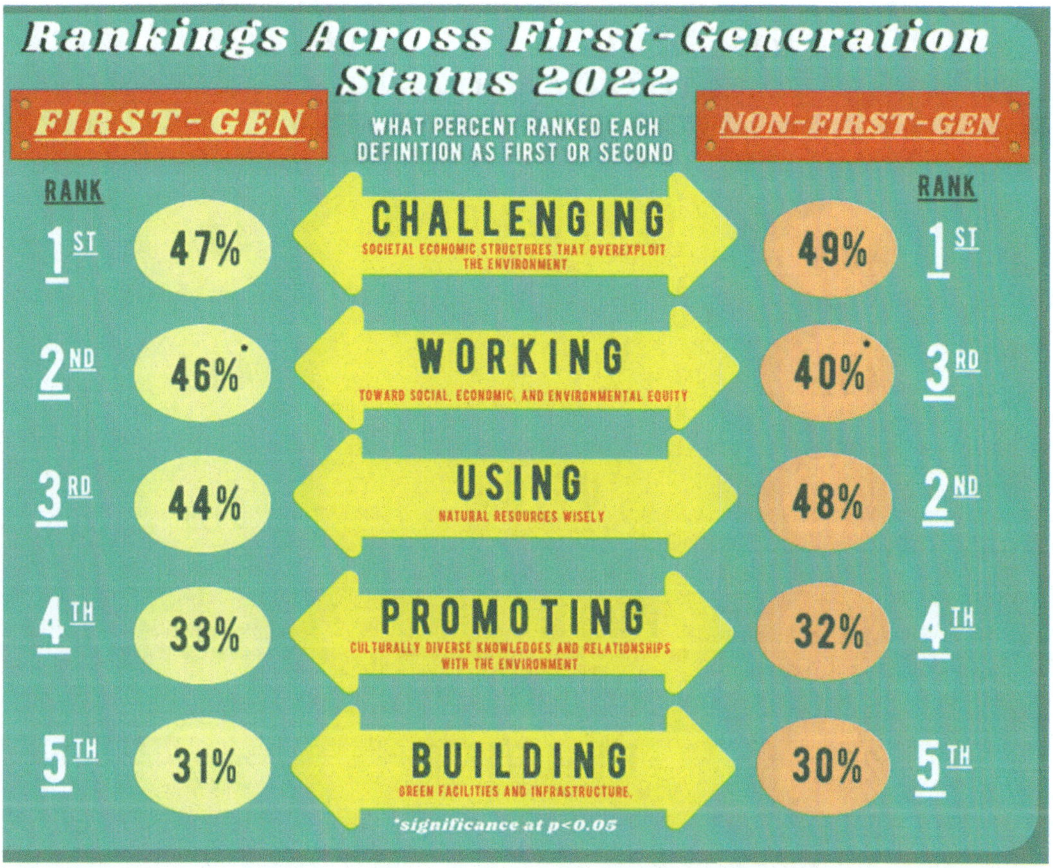

Fig. 4.8 Comparison of rankings of sustainability definitions by first-generation status in 2022

CHALLENGING, USING and WORKING sustainability definitions. Non-ENVS majors ranked them in descending order of importance, but each of these definitions were voted as either 1 or 2 in importance by over 40% of the respondents. We found no statistically significant differences between ENVS and non-ENVS majors.

Comparing 2019 and 2022, among ENVS majors, we do not find much difference: the same top three definitions received over 40%, with BUILDING the least popular. For non-ENVS students, across the different years, we found that USING was no longer the top definition, dropping to second after CHALLENGING, and that PROMOTING went up to fourth place, knocking BUILDING to fifth, the same as for ENVS students.

Discussion

The percentages reported in this chapter refer to the proportion of respondents who named a specific sustainability definition first or second place in importance out of five slots. If each definition was equally preferred, the breakdown would be 40% for each of the five options. Glancing at the tables, it is readily apparent that the results are not that far off from this 40% benchmark of parity. In other words, there is a plurality of understandings of sustainability among our undergraduate respondents. What this study reveals is that not only are there multiple conceptions of sustainability, but that each of the five possible dimensions we tested resonated with a substantial number of students.

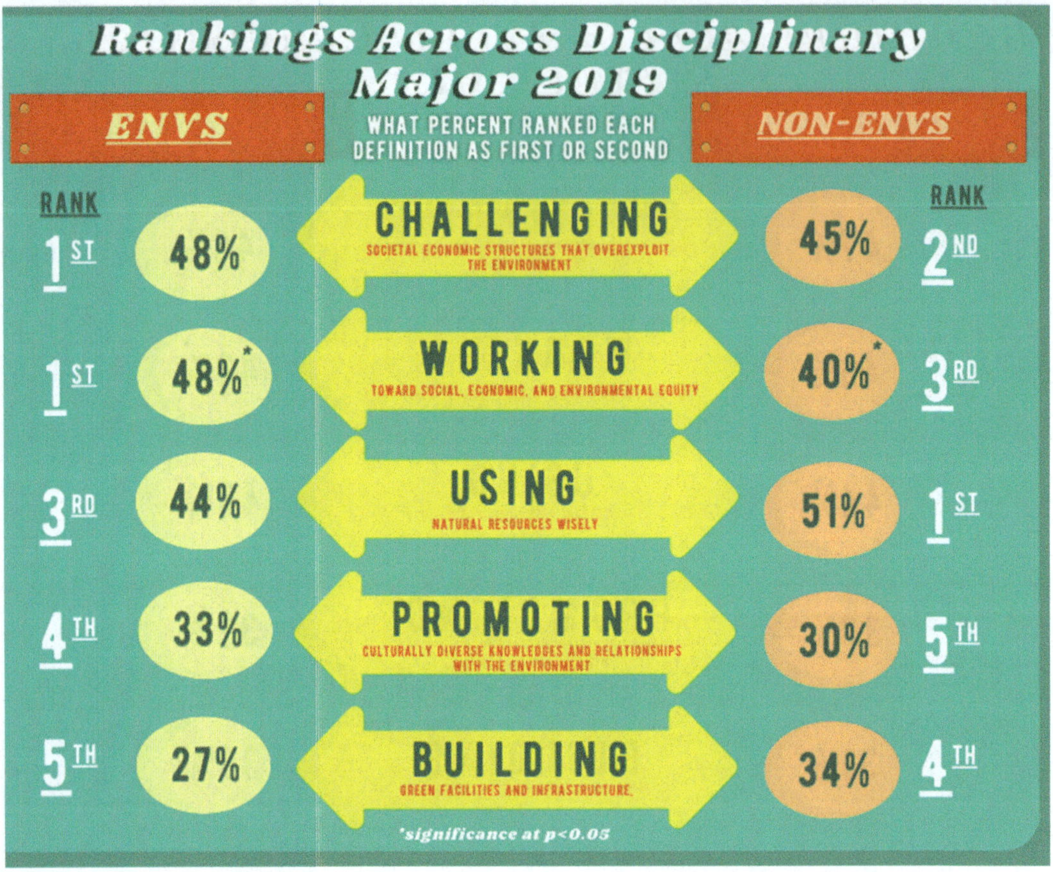

Fig. 4.9 Comparison of rankings of sustainability definitions by disciplinary major, namely Environmental Studies versus non-Environmental Studies, in 2019

In addition, we compared the results of two surveys, implemented in 2019 and 2022, and disaggregated the results by race, gender, first-generation status, and major (Environmental Studies versus other majors). We found that not only is there a plurality of definitions, but that the definitions that students find as most important or resonant change over time and change depending on who the students are. One of the most notable differences between 2019 and 2022 is that the dominant definition of sustainability went from USING natural resources wisely to CHALLENGING societal economic structures that overexploit the environment—this is a shift from focusing on individual acts to advocating for structural change. The only grouping in our study for whom USING is still the top-ranked sustainability definition are males. Given that in 2019, more than 50% of WNH students, continuing/non-first generation students, and non-ENVS majors ranked USING as first or second choice, this shift away from this framing of sustainability is noteworthy.

We also examined how the positionalities of the respondents can influence their definitions of sustainability. Comparing students of color and WNH, we found that the former exhibits more parity in definitions of sustainability: for students of color, the first and last ranked choices differ by only around 15%, whereas for WNH students, it is double that. In 2019, PROMOTING culturally diverse knowledges and relationships with the environment was the lowest ranked definition for WNH, but in 2022, that was no longer the case. For PoC, BUILDING and PROMOTING are consistently the two least popular definitions, but

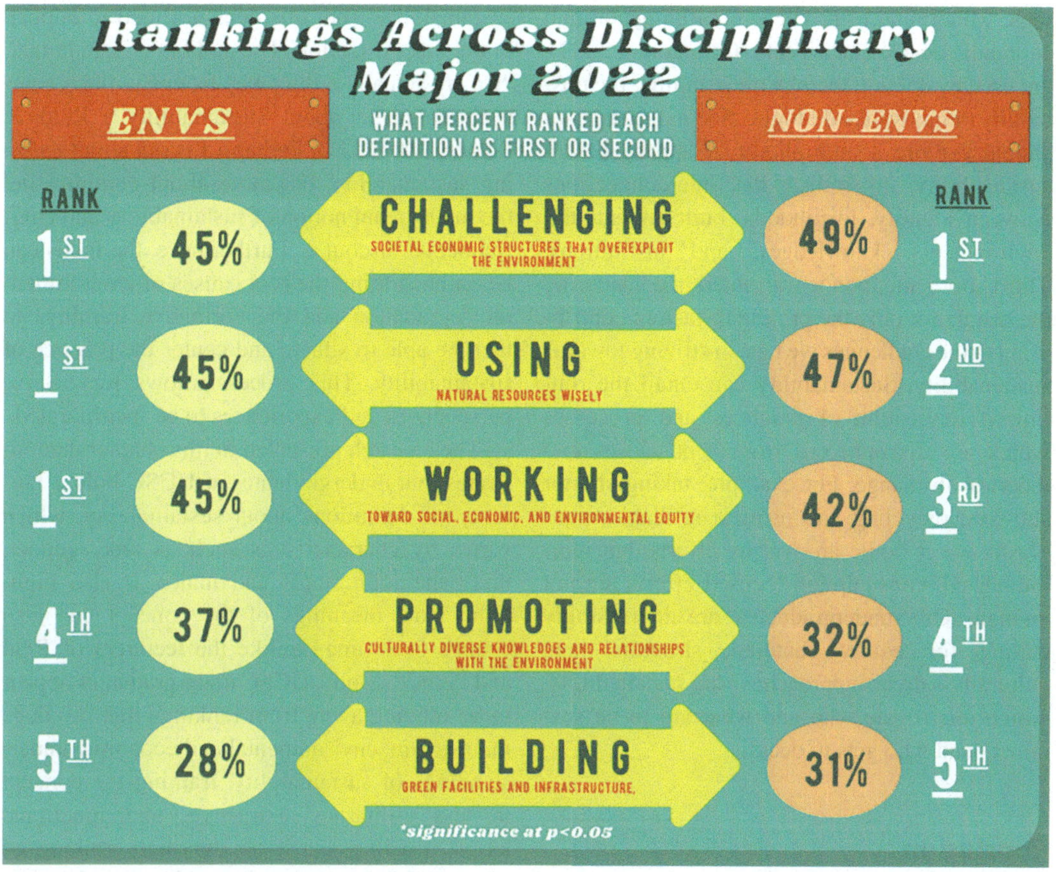

Fig. 4.10 Comparison of rankings of sustainability definitions by disciplinary major, namely Environmental Studies versus non-Environmental Studies, in 2022

in the recent survey, we examined more closely the differences between racial groups and found that for American Indian and Alaskan Native students, PROMOTING is by far the top ranked definition at 57%, over 20% higher than the other groups.

In terms of gender differences, the contrast between females and males became greater in the more recent survey. Female undergraduates rate CHALLENGING and WORKING higher than males, which reflect great concerns about structural change and equity, whereas males rate USING and BUILDING higher, which tend to be more dominant, mainstream sustainability framings. The rankings of non-binary students are similar to those of female students. In addition, our analysis revealed that, while in 2019, first-generation and continuing students were the same in their top three definitions, in 2022, first-generation students placed more emphasis on equity considerations in sustainability, ranking WORKING as second. Finally, in 2019, students majoring in ENVS prioritized structural economic and equity considerations in their rankings compared to other majors, but in 2022, there were no statistically significant differences between the two groups, reflecting a growing concern for environmental justice among our undergraduates more generally.

There are multiple implications of the findings in this chapter. First, it is important to be explicit in university discourse and policies to specify and operationalize what is meant by 'sustainability' and the intended goals and outcomes of any interventions. Leaving such conceptual clarification undone means that predominant understandings, which tend to privilege environmental and economic over social aspects (Colantonio, 2009), are used by default, thereby undermining the possibility of innovative dialogue. Second, campuses should explore a multiplicity of approaches to sustainability, to reflect the interconnections between identity, livelihoods, built and natural environments, knowledges, and institutions. Third, more attention is warranted in considering the distal, not only the proximal, factors contributing to the challenges we face in striving towards sustainability. Sustainability can entail the routinized replication of practices and processes which are disconnected from root causes and theories of change. For example, taking shorter showers, recycling and using a refillable, non-plastic water bottle are worthy efforts, but individualized, consumption-focused efforts are not enough. Our undergraduates are increasingly calling for equity centered, structural and culturally-salient approaches to sustainability, which interrogate who and what are to be sustained, and who gets to decide.

Conclusion

In a study of lecturers' understanding of sustainability, Reid and Petocz (2006) found that resource-centered definitions of sustainability were the most common, whether people were talking about material resources (e.g., minerals, water), biological resources (e.g., fish, crops) or

human resources (e.g., populations, economies). Some lecturers focused on basic definitions of sustainability (i.e., keeping something going), whereas others have a more justice-centered conception of sustainability, exemplified by the quote, "social justice, economic equity and cultural diversity...are really critically important elements [in the pursuit of ecological sustainability]" (Reid & Petocz, 2006: 117). These understandings are nested: someone with a justice-centered view can be aware of and use the resources view and articulate basic definitions of sustainability—but this does not tend to happen in the other direction. The authors conclude that, "we are actually a long way from sharing a common language about sustainability" (Reid & Petocz, 2006: 121). Perhaps it is not about defining sustainability, but more about complicating the mainstream notion of sustainability and environmental discourse. Furthermore, the focus can be on unearthing the root causes of environmental degradation and the normative framings to then be able to situate and center the practice of sustainability. This process allows for diverse perspectives and experiences to be incorporated.

The research presented in this chapter demonstrates that undergraduates at UCSC hold a plurality of definitions about sustainability, which varies by characteristics such as race, gender, first-generation status and major. It also highlights how meanings of the concept can also change over time. Unlike the lecturers of Reid and Petocz's study, UCSC undergraduates appear to be moving away from rankings that privilege mainstream, environmental and economic understandings of sustainability, framing the concept in less utilitarian, extractive, and functional terms. Findings such as the high ranking of PROMOTING as a definition of sustainability by Indigenous students reflect understandings of sustainability situated in culture, race, and history (i.e., settler colonialism, racial capitalism and white supremacy). As one undergraduate, an intern with PoCSC, aptly stated:

> Sustainability means different things to different people and it's important to recognize why those differences are there and what has shaped those definitions. It's important to make sure everyone from different backgrounds has a say in the sustainability agenda to make it equitable and allow it to connect with people...all definitions are important to sustainability and must be put into effect all together. The way we discuss and implement sustainability must be effective in all aspects and encapsulate all definitions at once. Our sustainability agenda on campus must be collaborative, because we all have different contributions and can play different roles in this fight. We must be open to learning from one another.

References

Artelle, K. A., Zurba, M., Bhattacharyya, J., Chan, D. E., Brown, K., Housty, J., & Moola, F. (2019). Supporting resurgent indigenous-led governance: A nascent mechanism for just and effective conservation. *Biological Conservation, 240*, 108284.

Azapagic, A., Perdan, S., & Shallcross, D. (2005). How much do engineering students know about sustainable development? The findings of an international survey and possible implications for the engineering curriculum. *European Journal of Engineering Education, 30*(1), 1–19.

Bajracharya, S. M., & Maskey, V. (2016). Students' awareness, values, perceptions, and behaviors toward environmental sustainability (ES): A comparative study. *International Journal of Sustainability Education, 12*(3), 1–14.

Balée, W. (2013). *Cultural forests of the Amazon: A historical ecology of people and their landscapes*. University of Alabama Press.

Ban, N. C., Frid, A., Reid, M., Edgar, B., Shaw, D., & Siwallace, P. (2018). Incorporate indigenous perspectives for impactful research and effective management. *Nature Ecology and Evolution, 2*(11), 1680–1683.

Berkes, F. (2018). *Sacred ecology* (4th ed.). Routledge.

Colantonio, A. (2009). Social sustainability: A review and critique of traditional versus emerging themes and assessment methods. In M. Horner, A. Price, J. Bebbington, & R. Emmanuel (Eds.), *Sue-mot conference 2009: Second international conference on whole life urban sustainability and its assessment: Conference proceedings* (pp. 865–885). Loughborough University.

Denevan, W. M. (1992). The pristine myth: The landscape of the Americas in 1492. *Annals of the Association of American Geographers, 82*(3), 369–385.

Dillon, L., & Sze, J. (2018). Equality in the air we breathe: Police violence, pollution, and politics of sustainability. In J. Sze (Ed.), *Sustainability: Approaches to environmental justice and social power* (pp. 246–270). New York University Press.

Emanuel, R., & Adams, J. N. (2011). College students' perceptions of campus sustainability. *International Journal of Sustainability in Higher Education, 12*(1), 79–92.

Environmental Protection Agency. (2014). *Enhancing sustainable communities with green infrastructure*. Retrieved August 11, 2021, from: https://www.epa.gov/sites/default/files/2016-08/documents/green-infrastructure.pdf

Escobar, A. (2010). Latin America at a crossroads. *Cultural Studies, 24*, 1–65.

Escobar, A. (2015). Degrowth, postdevelopment, and transitions: A preliminary conversation. *Sustainability Science, 10*, 451–462.

Fa, J. E., Watson, J. E. M., Leiper, I., Potapov, P., Evans, T. D., Burgess, N. D., Molnár, Z., Álvaro Fernández-Llamazares, A., Duncan, T., Wang, S., Austin, B. J., Jonas, H., Robinson, C. J., Malmer, P., Zander, K. K., Jackson, M. V., Ellis, E., Brondizio, E. S., & Garnett, S. T. (2020). Importance of indigenous peoples' lands for the conservation of intact forest landscapes. *Frontiers in Ecology and the Environment, 18*(3), 135–140. https://doi.org/10.1002/fee.2148

Feola, G. (2020). Capitalism in sustainability transitions research: Time for a critical turn? *Environmental Innovation and Societal Transitions, 35*, 241–250.

Harvey, D. (2006). *The limits to capital*. Verso.

Kagawa, F. (2007). Dissonance in students' perceptions of sustainable development and sustainability: Implications for curriculum change. *International Journal of Sustainability in Higher Education, 8*(3), 317–338.

Leach, M., Scoones, I., & Stirling, A. (2010). *Dynamic sustainabilities: Technology, environment, social justice*. Earthscan.

Lu, F., Valdivia, G., & Silva, N. L. (2017). *Oil, revolution and indigenous citizenship in Ecuadorian Amazonia*. Palgrave.

Maffi, L. (2005). Linguistic, cultural, and biological diversity. *Annual Review of Anthropology, 29*, 599–617.

Norman, E. S. (2017). Standing up for inherent rights: The role of indigenous-led activism in protecting sacred waters and ways of life. *Society and Natural Resources, 30*, 537–553.

Pellow, D. N. (2018). *What is critical environmental justice?* Polity.

Reid, A., & Petocz, P. (2006). University lecturers' understanding of sustainability. *Higher Education, 51*, 105–123.

Rose, J., & Cachelin, A. (2018). Critical sustainability: Incorporating critical theories into contested sustainabilities. *Journal of Environmental Studies and Sciences, 8*, 518–525. https://doi.org/10.1007/s13412-018-0502-9

Scoones, I. (2007). Sustainability. *Development in Practice, 17*(4–5), 589–596.

Summers, M., Corney, G., & Ghilds, A. (2004). Student teachers' conceptions of sustainable development: The starting-points of geographers and scientists. *Educational Research, 46*(2), 163–182.

Sze, J., Rademacher, A., Beamish, T., Grandia, L., London, J., Warren, L., Rose Middleton, B., & Ziser, M. (2018). Introduction. In J. Sze (Ed.), *Sustainability: Approaches to environmental justice and social power* (pp. 1–25). New York University Press.

Whitehead, M. (2014). Sustainability. In C. Death (Ed.), *Critical environmental politics: Interventions* (pp. 257–266). Routledge.

Whyte, K. P., & Cuomo, C. J. (2016). Ethics of caring in environmental ethics: Indigenous and feminist philosophies. In S. M. Gardiner & A. Thompson (Eds.), *The Oxford handbook of environmental ethics* (pp. 1–29). Oxford University Press.

Wilbur, T. G., & Roscigno, V. J. (2016). First-generation disadvantage and college enrollment/completion. *Socius, 2*. https://doi.org/10.1177/2378023116664351

Flora Lu is a Professor of Environmental Studies at the University of California, Santa Cruz and an ecological anthropologist specializing in conservation politics, environmental justice, political ecology and Indigenous resource governance. She served as the Provost of College Nine and John R. Lewis College from 2014 to 2023, during which time she helped establish the People of Color Sustainability Collective (PoCSC) and the Apprenticeship in Community Engaged Research (H)ACER. She was a Postdoctoral Fellow at Stanford University and an Assistant Professor in Ecology at the University of North Carolina at Chapel Hill before joining the UCSC faculty in 2008.

Tashina J. Vavuris is a Ph.D. student in the Environmental Studies Department at the University of California, Santa Cruz. She received her Master's degree in Sociology from San Diego State University, where she examined critical pedagogies and community-based practices. Her current research interests relate to environmental justice education with the intersection of critical pedagogies, participatory approaches, and Indigenous science relating to environmental sense-making and praxis.

Randy Uang is a Senior Survey Research Analyst in Institutional Research, Assessment, and Policy Studies at the University of California, Santa Cruz. He has experience conducting research on public health, public security, and environmental issues. He has a Ph.D. in Government from the University of Texas at Austin.

Cassandra Wood is a graduate from University of California, Santa Cruz in Environmental Studies and is currently in the Master's degree program in the Urban and Regional Planning department at University of California, Los Angeles. Their research lies in socio-spatial injustice in the built environment.

Anna Sher is a Director of Assessment and Survey Research at the University of California, Santa Cruz. She has been involved in survey research on various issues ranging from sustainability to diversity, equity and inclusion in all aspects of student experience and strategic planning for the university. She holds a Ph.D. in Sociology from Stony Brook University.

The Environmental Belief Paradox

5

Flora Lu, Randy Uang, and Anna Sher

Abstract

This chapter addresses the widespread underestimation of environmental concerns of low-income and people of color by people of all races and class positions. In the US, the archetype of the 'environmentalist' is associated with a white person, often a male, and someone of economic means. This pervasive notion is held by both higher-income whites, who believe that their demographic is the most concerned about sustainability, and low-income people of color, who underestimate the environmental attitudes of the groups to which they belong. Research at the national level shows that this perception, called the "environmental belief paradox", is not supported by evidence and perpetuates the discounting of minoritized groups' sustainability concerns. We engage the concept of the environmental belief paradox and explore how levels of concern about various environmental issues vary by race, gender, first-generation status and major. In examining survey data from 2019 and 2022 from UCSC undergraduates, we found that race and class (approximated by first-generation status) provide less explanatory power than gender and major when it comes to importance placed on environmental issues.

F. Lu (✉) · R. Uang · A. Sher
University of California, Santa Cruz, CA, USA
e-mail: floralu@ucsc.edu

Keywords

Environmental belief paradox · Environmental concern · Race · Gender · First-generation status

Introduction

> I feel that white middle/upper class students are always the ones most interested in the environmental movement and it's not that other students are not interested, I think they might not feel as comfortable in that environment or it might not be close to their interests. (Male, Senior, Sociology Major, white)

> Almost all of the ENVS (Environmental Studies) major is comprised of [sic] White students... I believe this is because White populations (often times more wealthy) have access to environmental education at younger ages... I think the least interested population of students would be African American, Latino and Asian. (Female, Sophomore, ENVS major, South Asian/white)

> I feel like the lower income/minorities are a lot less interested in environmental studies because their culture/background does not advocate for protecting the environment. (Male, Junior, ENVS/ECON major, Latinx)

> I think people who don't know how their actions impact the environment are way less likely to become involved in the movement. People who are

F. Lu, E. Murai (eds.), *Critical Campus Sustainabilities*, Sustainable Development Goals Series,
https://doi.org/10.1007/978-3-031-30929-8_5

> ignorant to the consequences of their actions are less likely to become involved in trying to change them and the environment which bears the brunt of their actions. (Female, Junior, Politics major, white)

> Those least interested in environmental issues at UCSC were those who have never thought about sustainability as it relates to their lives…those who have grown up in households where disposable products are used without thought; those who see sustainability as an inconvenience to their way of life. (Female, Junior, Literature major, white)

These quotes are taken from the two dozen semi-structured interviews conducted by Chelsea Pack, an Environmental Studies major at the University of California, Santa Cruz (UCSC) who wrote a senior thesis entitled, "Privileged perspectives or diverse discourse? Disparities in student perceptions of and participation in the environmental movement at UCSC" (Pack, 2014). The comments highlight a widespread belief that white, middle-class people are environmentalists, while racial and ethnic minorities and low-income groups are least concerned by environmental issues. A lack of environmental concern or awareness, in turn, is judged to be akin to a character flaw, a lapse in one's socialization or upbringing. These UCSC students' interview responses suggest that racism and classism pervade environmental discourse at this university, disrupting the idealistic view that UCSC is an eco-Mecca where constituents have managed to transcend an environmental history fraught with conflict and inequality to unite over a shared passion for the planet. In order for institutions of higher education to produce environmentally literate members of society, it is necessary that they incorporate a range of perspectives, values, goals, and environmental understandings. Considerations of power, race, class and gender are crucial to students' ability to think critically and holistically about environmental issues.

The student quotes highlight what Pearson et al. call the "environmental belief paradox," which the authors define as "a tendency to misperceive groups that are among the most environmentally concerned and most vulnerable to a wide range of environmental impacts as least concerned about the environment" (Pearson et al., 2018: 12430). This pervasive notion is held by both higher-income whites, who believe that their demographic is the most concerned about sustainability, and low-income people of color, who underestimate the environmental attitudes of the groups to which they belong. Pearson et al. (2018) utilized a probability-based sample of US adults in which they asked participants to self-report their level of environmental concern and whether they identified as an environmentalist. They then asked respondents to rate the level of environmental concern of various demographic groups and also describe their conception of an 'environmentalist', based on age, education, race, and class.

The authors found that white respondents' self-reported environmental concern was much lower than what was ascribed to them, and people of color reported significantly higher levels of concern, on average, than whites (Pearson et al., 2018: 12431). Respondents also underestimated the environmental concern of poor Americans to a greater extent than that of wealthy Americans (Pearson et al., 2018). The gap between public perception and self-reported concern was highest for Latinx and poor people, with Asians and Blacks closely behind. Blacks, Latinx, and Asians all significantly underestimated their own racial/ethnic group's level of environmental concern compared to the aggregate of self-reporting among those in each of these groups. People of color also misperceive their personal level of environmental concern as greater than that of their ingroup, a phenomenon called "pluralistic ignorance."

In particular, white males significantly underestimated the environmental concern of all target groups relative to the mean self-reported concern of each group. In addition, white males reported less environmental concern than all other respondents taken together. White males were the only race-gender group for whom an inverse pluralistic ignorance effect was found, i.e., in which white males reported being less concerned about the environment than others in their ingroup. In other words, white males were the only race/gender category in which individuals rated their own personal environmental concern as lower than the ratings of how they think others in their group feel. White females reported less environmental concern than non-Whites of either gender, and there was no gender difference in environmental concern among non-Whites (Pearson et al., 2018).

There was a broad consensus as to who constitutes an environmentalist: participants of all racial, ethnic and class backgrounds surveyed described an environmentalist as white, wealthy, highly educated, and younger in age. Pearson et al. (2018) found that both white and non-white respondents *disassociated* non-whites from environmentalists. These findings are not surprising given the way representations of outdoor recreation, like mountain climbing, snowboarding and mountain biking, what Braun (2003) refers to as elements of "risk culture," are highly racialized as white. Braun (2003) goes on to argue that contemporary forms of risk culture are an extension of Western frontier imaginaries, in which Euro-American settlers go forth to explore, conquer and tame presumably unsettled wild and primitive lands, thus creating a continued narrative that equates recreational nature with environmentalism and whiteness, and one in which non-whites have no place. Yet this stereotype does not match participants' self-reports as environmentalists, thus upending the narrative: while 50.2% of white respondents self-identified as an environmentalist, 65.2% of Latinos and 67.6% of Asians did. The authors also found that neither education nor income was a significant predictor of self-identification as an environmentalist (Pearson et al., 2018: 12431).

The implications of these results are important. People are motivated by norms to engage in pro-environmental behavior, and this is especially the case for minoritized and lower-income groups. Citing literature from social psychology, Pearson et al. (2018: 12430) state that these marginalized groups "are particularly attuned to norms signaling what is appropriate and preferred by their ingroup and, consequently, are more likely to engage in behaviors perceived as normative or 'identity-congruent'." So if communities of color are perceived as not being interested in environmental concerns, normative and perceptual barriers are erected, inhibiting involvement in environmental action. At UCSC, for example, this phenomenon has manifested in African American students sharing that they were discouraged from participating in campus environmental efforts or even talking about environmental concerns with other Black peers because this was deemed "acting white." The prevalent stereotype at UCSC of an environmentalist as a white "hippie" able to afford expensive clothing, outdoor recreation gear and local, organic food can alienate low-income students of color and inhibit people from affiliating with environmental groups and working towards environmental causes.

In this chapter, we explore whether and how the environmental belief paradox manifests on a college campus, focusing specifically on comparing people's self-reported concerns about environmental issues. We differentiate between various environmental concerns, from those focused on infrastructure and resource use, to environmental justice issues, and ask the following:

1. How do the self-reported environmental concerns of white, non-Hispanic undergraduates compare with those of students of color?
2. How do the various racial/ethnic groups compare and contrast in terms of their self-reported environmental concerns?
3. Beyond race and ethnicity, how do undergraduates differ in their environmental concerns based on gender, major, and first-generation status?

Methods

The People of Color Sustainability Collective (PoCSC) administered surveys in Winter 2019 and Winter/Spring 2022 to understand undergraduate student perspectives on sustainability and environmental aspects of campus. All enrolled students (16,983 undergraduates in 2019, and 17,064 in 2022) were invited to take an online survey where they were asked to share their perceptions of and participation in the environmental sustainability movement on campus. The survey response rates were 15% (n = 2546) in 2019 and 13% (n = 2273) in 2022. Survey respondents were representative of the student population in terms of race and ethnicity, first generation status, transfer status, and class level. Weighted data was used for our analysis to ensure proper representation of the student population in terms of gender and ethnicity. In addition to the survey data, the analyses incorporated institutional data on students' demographic background (gender, race/ethnicity, first generation status) and academic path (major, college, class level, and transfer status). The authors worked closely with UCSC's Institutional Research, Assessment and Policy Studies (IRAPS) office with survey design, administration, and data analysis.The survey was approved by the Institutional Review Board.

This chapter focuses on survey results that asked students to think about their personal level of importance (not important to very important) of environmentally related issues. The questions were:

Please rate your personal level of importance of each of the following issues (Not important, Somewhat Important, Important, Very Important):

- **Environmental Health** (address chemical and toxic exposures in air, water, soil and food in natural and built environment)
- **Agroecology** (promote the productivity, stability and equitability of agroecosystems through mimicking natural processes)
- **Conservation of Natural Resources & Protection of Biodiversity and Habitats**
- **Food Security** (access to healthy and affordable food as a basic need)
- **Food Sovereignty** (food as central to cultural identity as well as nourishment, such as revitalizing native and local foodways)
- **Environmental Justice** (intersection of race, class and power in environmental policy making and distribution of burdens of environmental degradation)
- **Climate Change** (human-caused, disruptive shifts in climate patterns regionally and globally)
- **Non-Fossil Fuel, Renewable Energy**
- **Green Building Design**
- **Shared, Accessible, and Low-Emission Transportation**
- **Waste Reduction & Prevention**
- **Water Conservation**

The research team utilized descriptive statistics to characterize what proportion of these student population groups ranked environmental issues as important/not important, and then ran statistical tests (chi-square test for dichotomous categorical variables) to determine if the differences were statistically significant. Note that in some cases where differences are statistically significant, they can be small in substance, capturing differences of a few percentage points.

Findings

PoC vs. WNH

We first address the first question, how do the self-reported environmental concerns of white, non-Hispanic (WNH) undergraduates compare with those of students of color (PoC)? We start by discussing the results for 2019 and then 2022. At first glance, a main take-away from Table 5.1 is that all students deem environmental issues of great importance. Over 90% of students felt that over half of the issues listed were important or very important. Of the dozen issues listed, green building design had the lowest percentage of students rating it as important or very important, and climate change was the most important issue to undergraduates.

First, there are several issues for which there was no difference in ratings by WNH and PoC students: waste reduction and prevention, sustainable water use, food security, and agroecology. At $p < 0.05$, two issues, environmental justice and climate change were found to be statistically significant, with slightly more PoC rating the former as important and slightly more WNH students rating the latter as important. At $p < 0.01$, a higher percentage of WNH rated sustainable transportation, green building design, and conservation as important or very important. At $p < 0.001$, we see the largest differences between the two groups of students, with more WNH students rating renewable energy as important compared to PoC, and more PoC valuing food sovereignty than WNH.

Table 5.2 shares the results for 2022. The percentage of PoC and WNH reporting environmental issues as important/very important have increased across the board. The most statistically significant contrasts between these two categories are the findings that WNH are more concerned with agroecology, climate change and renewable energy, and PoC with food sovereignty.

Race/Ethnicity

The category of PoC overgeneralizes a great deal of heterogeneity. In this section, we ask, how do the various racial/ethnic groups compare and contrast in terms of their self-reported environmental concerns? As summarized in Table 5.3, the results are complicated. From the outset it is helpful to note that these results are influenced by differences in population size of different racial and ethnic groups on campus. In Winter 2019, when the survey was implemented, the UCSC undergraduate population's racial/ethnic composition was as follows: Asian-American and Pacific Islander (28%); Hispanic/Latinx (27%); White, non-Hispanic (30%); International (8%); African American/Black (4%), and American Indian/Alaskan Native (1%). Pacific Islander students constituted less than 1% and the results for them are not shown due to small sample size. Note that some of the statistically significant differences, or lack of difference, can be attributed to disparities in relative sample sizes of the different groupings. This seems to be the case for climate change and renewable energy. The directionality of the results should be considered with that caveat in mind.

However, these data point to some interesting findings. Consistent with the previous section, white students perceive issues of environmental health, climate change, conservation, renewable energy, and green buildings to be important. Hispanic/Latinx students appear to be driving the concern for food sovereignty and agroecology among PoC. Although their population size is small at UCSC, the results for African American and Black (ABC) students reflect a valuing of

Table 5.1 Results of statistical analysis of PoC vs. WNH in 2019

Ranked "Important" or "Very Important"	PoC n = 1530	WNH n = 776	Level of significance
Environmental health	93%	95%	$p < 0.05$
Agroecology	79%	81%	–
Conservation of natural resources & protection of biodiversity and habitats	90%	94%	$p < 0.01$
Food security	95%	93%	–
Food sovereignty	78%	68%	$p < 0.001$
Environmental justice	85%	81%	$p < 0.05$
Climate change	95%	97%	$p < 0.05$
Non-fossil fuel, renewable energy	91%	96%	$p < 0.001$
Green building design	82%	87%	$p < 0.01$
Shared, accessible, and low-emission transportation	88%	92%	$p < 0.01$
Waste reduction & prevention	92%	93%	–
Water conservation	91%	90%	–

Table 5.2 Results of statistical analysis of PoC vs. WNH in 2022

Ranked "Important" or "Very Important"	PoC n = 1383	WNH n = 721	Level of significance
Environmental health	95%	97%	$p < 0.05$
Agroecology	76%	84%	$p < 0.001$
Conservation of natural resources & protection of biodiversity and habitats	93%	96%	$p < 0.01$
Food security	97%	96%	–
Food sovereignty	84%	75%	$p < 0.001$
Environmental justice	87%	89%	–
Climate change	95%	99%	$p < 0.001$
Non-fossil fuel, renewable energy	92%	97%	$p < 0.001$
Green building design	80%	85%	$p < 0.01$
Shared, accessible, and low-emission transportation	90%	93%	$p < 0.05$
Waste reduction & prevention	93%	95%	–
Water conservation	94%	93%	–

environmental issues, especially food security, climate change, waste reduction, environmental justice, and water conservation. The proportion of Asian American respondents ascribing high levels of importance to environmental concerns varied, and were generally in the mid-range (Table 5.3).

In 2022, we found that Hispanic/Latinx students reported high levels of concern for seven of the twelve environmental issues; and that ABC students similarly expressed concern for environmental justice and food sovereignty (Table 5.4).

Beyond race and ethnicity, how do undergraduates differ in the levels of importance they ascribe to environmental concerns based on gender, major, and first-generation status?

Gender

In Academic Year 2018–2019, the undergraduate student population at UC Santa Cruz was 48.4% female and 50.7% male. We found that across the board, women ranked higher levels of importance in all categories over men. As noted in Table 5.5 below, all categories are statistically significant at $\alpha < 0.001$. This is a resounding finding and points to gender as a key characteristic in terms of understanding environmental concern.

In Academic Year 2021–2022, the undergraduate student population at UC Santa Cruz was 45.9% female and 50.2% male and 3.9% non-binary/other. This was the first year that data were made available that went beyond the gender binary, a much needed and overdue advancement. Table 5.6 reports the results of statistical analysis comparing the responses of these groups. Women placed greater importance on six of the twelve issues than men, and non-binary students joined women in expressing more concern than men for issues of food sovereignty, environmental justice and transportation. For the issue of renewable energy, non-binary students were statistically higher than men in their ranking of this issue as important.

First Generation vs. Continuing Generation

In Academic Year 2018–2019, 38.1% of the undergraduate student population at UC Santa Cruz was first-generation, i.e., neither parent(s)/guardian(s) received a four-year college degree; 59.3% were non-first-generation, or continuing generation. We found that first generation college students ranked higher levels of importance in about half of the categories over continuing gen-

Table 5.3 Results of statistical analysis for various categories of race/ethnicity in 2019

Ranked "Important" or "Very Important"	African American/ Black (ABC) n = 107	American Indian/ Alaska Native (AIAN) n = 19	Asian American (AA) n = 719	Hispanic/ Latinx (LX) n = 676	White, non-Hispanic (WNH) n = 776	International (Intl) n = 197	Statistical significance ($p < .05$) in Chi-Squared tests
Environmental health	92%	85%	92%	95%	95%	85%	WNH, LX > Intl
Agroecology	80%	73%	74%	82%	81%	80%	LX > AA
Conservation	92%	79%	86%	94%	94%	86%	WNH, LX > AA, Intl
Food security	96%	90%	94%	96%	93%	93%	None
Food sovereignty	83%	63%	73%	83%	68%	72%	LX > AA, WNH, Intl
Environmental justice	93%	79%	80%	89%	81%	77%	ABC > WNH, AA, Intl
Climate change	96%	79%	93%	96%	97%	86%	WNH > AA, Intl, AIAN LX > Intl, AIAN AA > AIAN
Renewable energy	89%	90%	90%	92%	96%	79%	WNH > LX, AA, Intl LX > Intl AA > Intl
Green building design	87%	76%	79%	84%	87%	75%	WNH > AA, Intl
Transportation	92%	79%	86%	90%	92%	89%	WNH > AA
Waste reduction & prevention	94%	90%	92%	93%	93%	87%	None
Water conservation	93%	90%	89%	94%	90%	90%	LX > AA

Table 5.4 Results of statistical analysis for various categories of race/ethnicity in 2022

Ranked "Important" or "Very Important"	African American/ Black (ABC) n = 105	American Indian/ Alaska Native (AIAN) n = 12	Asian American (AA) n = 606	Hispanic/ Latinx (LX) n = 652	White, non-Hispanic (WNH) n = 721	International (Intl) n = 124	Statistical significance ($p < .05$) in Chi-square tests
Environmental health	93%	100%	92%	97%	97%	93%	LX > AA
Agroecology	72%	83%	72%	80%	84%	76%	LX > AA
Conservation	93%	100%	90%	95%	96%	82%	LX > AA
Food security	98%	100%	94%	98%	96%	94%	LX > AA
Food sovereignty	91%	88%	79%	88%	75%	80%	ABC, LX > AA
Environmental justice	96%	88%	83%	90%	89%	80%	ABC, LX > AA
Climate change	94%	100%	94%	95%	99%	87%	–
Renewable energy	88%	100%	90%	94%	97%	79%	LX, AA > ABC
Green building design	80%	76%	79%	81%	85%	75%	–
Transportation	91%	100%	89%	90%	93%	79%	–
Waste reduction & prevention	90%	100%	91%	95%	95%	92%	LX > AA, ABC
Water conservation	95%	82%	91%	96%	93%	86%	LX > AA, AIAN

Table 5.5 Results of statistical analysis by gender in 2019

Ranked "Important" or "Very Important"	Women n = 1206	Men n = 1285	Level of significance
Environmental health	95%	91%	$p < 0.001$
Agroecology	82%	76%	$p < 0.001$
Conservation of natural resources & protection of biodiversity and habitats	95%	87%	$p < 0.001$
Food security	97%	92%	$p < 0.001$
Food sovereignty	85%	63%	$p < 0.001$
Environmental justice	92%	74%	$p < 0.001$
Climate change	98%	92%	$p < 0.001$
Non-fossil fuel, renewable energy	94%	89%	$p < 0.001$
Green building design	87%	79%	$p < 0.001$
Shared, accessible, and low-emission transportation	93%	86%	$p < 0.001$
Waste reduction & prevention	95%	90%	$p < 0.001$
Water conservation	96%	86%	$p < 0.001$

eration college students, with the largest differences found in issues of environmental justice, food security and food sovereignty (Table 5.7).

In Academic Year 2021–2022, 34.8% of undergraduates were first-generation and 62.6% were continuing generation. Other than first-generation students having slightly higher percentages for food security and food sovereignty, there were no significant differences between the two groups (Table 5.8).

Environmental Majors vs. Other Majors

In 2019, we found that environmental majors (Environmental Studies (ENVS) and Environmental Sciences (ESCI)) reported higher levels of importance in all categories except food security, renewable energy and green buildings over all other majors. As seen below, some categories (namely agroecology, environmental justice, and water conservation) are more statistically significant than others (Table 5.9).

In 2022, we found that environmental majors expressed higher levels of importance for environmental issues than other majors, except for environmental health, food security, and renewable energy for which the two groups were the same (Table 5.10). This parity of concern was consistent for the latter two issues from 2019, but new for environmental health, which other majors placed more importance on in 2022. Other majors narrowed the gap of concern for environmental issues such as food sovereignty and water conservation, but for issues such as green buildings and waste reduction, environmental majors slightly increased the gap between them and other majors.

Discussion

The environmental belief paradox refers to the prevalent notion that environmental concern, advocacy and expertise is concentrated in certain subsets of society, namely white, male, wealthy, young, and formally educated people. The stereotype of an 'environmentalist' corresponds to dominant identities in our society, and conveys additional forms of power, authority and capital. Pearson et al. (2018) found that people of color underestimated the degree of environmental awareness and commitment to sustainability in their own communities. In this study, we tackle this key facet of the environmental belief paradox: what levels of concern are expressed by different groups around environmental issues? Where are the similarities and differences? How

Table 5.6 Results of statistical analysis by gender in 2022

Ranked "Important" or "Very Important"	Women n = 1044	Men n = 1139	Non-binary/genderqueer n = 69	Significant differences ($p < 0.05$)
Environmental health	96%	94%	97%	–
Agroecology	78%	78%	84%	–
Conservation of natural resources & protection of biodiversity and habitats	96%	91%	95%	Women > men
Food security	98%	95%	99%	Women > men
Food sovereignty	89%	73%	89%	Women, non-binary > men
Environmental justice	95%	80%	93%	Women, non-binary > men
Climate change	98%	94%	98%	Women > men
Non-fossil fuel, renewable energy	94%	92%	98%	Non-binary > men
Green building design	85%	78%	84%	Women > men
Shared, accessible, and low-emission transportation	93%	87%	98%	Non-binary, women > men
Waste reduction & prevention	95%	92%	95%	Women > men
Water conservation	96%	91%	92%	Women > men

Table 5.7 Results of statistical analysis by first-generation status in 2019

Ranked "Important" or "Very Important"	First generation n = 893	Continuing generation n = 1649	Level of significance
Environmental health	94%	93%	–
Agroecology	80%	79%	–
Conservation of natural resources & protection of biodiversity and habitats	93%	90%	$p < 0.01$
Food security	97%	93%	$p < 0.001$
Food sovereignty	83%	70%	$p < 0.001$
Environmental justice	90%	79%	$p < 0.001$
Climate change	95%	94%	–
Non-fossil fuel, renewable energy	92%	92%	–
Green building design	85%	82%	–
Shared, accessible, and low-emission transportation	90%	89%	–
Waste reduction & prevention	94%	91%	$p < 0.01$
Water conservation	93%	90%	–

Table 5.8 Results of statistical analysis by first-generation status in 2022

Ranked "Important" or "Very Important"	First generation n = 740	Continuing generation n = 1528	Level of significance
Environmental health	95%	95%	–
Agroecology	80%	78%	–
Conservation of natural resources & protection of biodiversity and habitats	93%	93%	–
Food security	98%	96%	$p < 0.05$
Food sovereignty	83%	80%	$p < 0.05$
Environmental justice	88%	87%	–
Climate change	95%	96%	–
Non-fossil fuel, renewable energy	93%	93%	–
Green building design	82%	81%	–
Shared, accessible, and low-emission transportation	92%	90%	
Waste reduction & prevention	95%	93%	–
Water conservation	94%	93%	–

consistent are these findings over time, comparing 2019 and 2022? We disaggregate the category of environmental concern to articulate a range of issues, from buildings and transportation, to protecting water/land/ecosystems, to biophysical and cultural sustenance. Moreover, we explore how environmental concerns vary not just by race and ethnicity, but also gender, first generation status, and major.

In line with Pearson et al. (2018), our results for UCSC undergraduates also demonstrate that students of color are very concerned about environmental issues, similar to their white counterparts. On the issue of food sovereignty, PoC are higher than WNH, and WNH are higher on certain, mainstream environmental concerns such as conservation and renewable energy. Analyses disaggregating the category of PoC into African American, Black and Caribbean; Asian American; Hispanic/Latinx; and American Indian/Alaskan Native highlight the heterogeneity elided by the category of PoC, reflect limitations in sample sizes for certain groups, and call for further research to better understand patterns of environmental concern.

Gender emerged as a more robust lens through which to understand environmental concern. In 2019, a greater percentage of women expressed

Table 5.9 Results of statistical analysis by major in 2019

Ranked "Important" or "Very Important"	ENVS/ ESCI majors n = 222	Other majors n = 2324	Level of significance
Environmental health	99%	93%	$p < 0.01$
Agroecology	92%	78%	$p < 0.001$
Conservation of natural resources & protection of biodiversity and habitats	97%	90%	$p < 0.01$
Food security	96%	94%	–
Food sovereignty	82%	74%	$p < 0.01$
Environmental justice	93%	82%	$p < 0.001$
Climate change	98%	94%	$p < 0.05$
Non-fossil fuel, renewable energy	95%	91%	–
Green building design	87%	82%	–
Shared, accessible, and low-emission transportation	95%	89%	$p < 0.01$
Waste reduction & prevention	97%	92%	$p < 0.01$
Water conservation	98%	90%	$p < 0.001$

Table 5.10 Results of statistical analysis by major in 2022

Ranked "Important" or "Very Important"	ENVS/ ESCI majors n = 191	Other majors n = 2078	Level of significance
Environmental health	97%	95%	–
Agroecology	93%	77%	$p < 0.001$
Conservation of natural resources & protection of biodiversity and habitats	99%	93%	$p < 0.001$
Food security	96%	96%	–
Food sovereignty	88%	80%	$p < 0.01$
Environmental Justice	99%	86%	$p < 0.001$
Climate change	98%	95%	$p < 0.05$
Non-fossil fuel, renewable energy	96%	93%	–
Green building design	88%	81%	$p < 0.05$
Shared, accessible, and low-emission transportation	96%	90%	$p < 0.01$
Waste reduction & prevention	100%	93%	$p < 0.001$
Water conservation	98%	93%	$p < 0.01$

environmental concern for all issues surveyed, a finding consistent with other research (e.g., Zelenzy et al., 2000). This pattern has been characterized as the "eco-gender gap", portraying women as caregivers in the home and of the planet, depicted as inherently more prosocial, altruistic and empathetic. The gendering of environmental concern signifies that actions such as being vegan or using a reusable shopping bag are portrayed as feminine (Hunt, 2020). Past male environmental icons such as John Muir were associated with wilderness preservation, preserving spaces for rugged recreation by elite, privileged men (Cronon, 1996). Some of the highest-profile female environmental leaders, especially those who are women of color or youth, are the target of misogyny, often by older, conservative men who masculinize the modern, industrial society and feel it is under threat (Anshelm & Hultman, 2014; Gelin, 2019). In an overdue recognition that gender is not binary, in 2022, our analysis included non-binary and genderqueer people. We found that women and/or non-binary people reported higher environmental concern than men for all but two issues. Gendered (and intersectional) analysis of environmental concern is an important direction for future research and curricular offerings at UC Santa Cruz.

In 2019, the survey results indicated that more first-generation students than continuing students expressed concern about food security, food sovereignty and environmental justice, and to a lesser degree conservation and waste reduction. By 2022, however, despite slightly higher importance placed on food security and food sovereignty, there were no differences between first-generation and continuing students' environmental concerns, attributable to the higher percentages reported by the latter.

Previous research (Lu et al., 2018) found that one of the most powerful factors associated with student participation in the environmental sustainability movement at UCSC was majoring in Environmental Studies. These majors were 5.2 times more likely to participate compared to all other students. The data collected here confirms these trends: in 2019, we found that ENVS majors placed more importance than non-ENVS majors on three-fourths of the environmental issues queried. In 2022, ENVS majors reported higher percentages of environmental concern for two-thirds of the issues queried, and of those, almost half were significant at $p < 0.001$. These issues corresponded to key foci of the department, such as agroecology, conservation, and environmental justice, as well as waste reduction, water conservation and shared, accessible, and low emission transportation. The percentages reported by ENVS majors in 2022 are the highest of any group in the two surveys, which, while impressive, is perhaps not surprising given that these students are dedicating their academic work to the environmental field.

Conclusion

As the student quotes that began the chapter allude to, the populations who bear the brunt of ecological degradation are largely perceived as not caring enough about, not knowing enough about, or—more insidiously—not culturally predisposed towards environmentalism. They discursively reflect educational deficit thinking (Valencia, 2010) and cultural deprivation theory, or the idea that student deficiencies are the result of an impoverished culture. Utilizing "language couched in charitable concern," such an approach in addressing a lack of environmental awareness (or other problems) focuses on "repairing the individual student and as a consequence, the larger societal structures that maintain inequity, including institutional norms, values, and assumptions, remain unquestioned" (Castro, 2014: 409).

One of these assumptions, held not only among undergraduate students but the broader society, is that those who already hold positions of relative privilege and power are also the ones who know best and care most about environmental concerns. The environmental belief paradox, described by Pearson et al. (2018), impedes efforts to address environmental inequities by dismissing and undermining the concerns and engagement that underrepresented groups and minoritized populations have with environmental issues. Pearson et al. (2018: 12429) not only demonstrate that this pattern of erroneous attributions and beliefs is "largely invariant across demographic groups," but also provide a promising way forward to correct this misperception. Using a randomized experiment, they were able to bridge the perceived gap between whites' and nonwhites' environmental concerns through exposure to a racially and ethnically diverse environmental organization. Namely, "the more inclusive and welcoming the organization was perceived to be, the more participants expressed interest in participating in the organization and the more favorable their attitude was toward environmental organizations, generally," especially for non-white respondents (Pearson et al., 2018: appendix).

In early November of 2015, in one of its first campus programs, PoCSC hosted a discussion entitled "Re-Imagining the Environmentalist." In a crowded room, students, staff and faculty of diverse backgrounds and identities broke into small groups to challenge the common archetype of the environmentalist as a wealthy, formally educated, young, white male. In creating this space, and many other spaces in which the environmental belief paradox was challenged, troubled, and upended, PoCSC has intentionally sought to welcome, include, support, and amplify the experiences, concerns, aspirations and expertise of those who have been historically marginalized not only as environmental leaders but as integral members of society.

References

Anshelm, J., & Hultman, M. (2014). A green fatwa? Climate change as a threat to the masculinity of industrial modernity. *Normat, 9*(2), 84–96. https://doi.org/10.1080/18902138.2014.908627

Braun, B. (2003). 'On the raggedy edge of risk': Articulations of race and nature after biology. In D. Moore, J. Kosek, & A. Pandian (Eds.), *Race, nature, and the politics of difference* (pp. 175–203). Duke University Press.

Casto, E. L. (2014). "Underprepared" and "at-risk": Disrupting deficit discourses in undergraduate STEM recruitment and retention programming. *Journal of Student Affairs Research and Practice, 51*(4), 407–419.

Cronon, W. (1996). The trouble with wilderness; or, getting Back to the wrong nature. In W. Cronon (Ed.), *Uncommon ground: Rethinking the human place in nature* (pp. 69–90). W.W. Norton & Company.

Gelin, M. (2019, August 28). The misogyny of climate deniers. *The New Republic.* https://newrepublic.com/article/154879/misogyny-climate-deniers.

Hunt, E. (2020, February 6). The eco-gender gap: Why is saving the planet seen as women's work? *The Guardian* https://www.theguardian.com/environment/2020/feb/06/eco-gender-gap-why-saving-planet-seen-womens-work

Lu, F., Hernandez Rosser, R., Renteria, A., Kim, N., Erickson, E., Sher, A., & O'Connor, L. (2018). Inclusive sustainability: Environmental justice in higher education. In W. L. Filho, J. Callewaert, & R. Marans (Eds.), *Handbook of sustainability and social science research* (pp. 63–81). Springer.

Pack, C. (2014). *Privileged perspectives or diverse discourse? Disparities in student perceptions of and participation in the environmental movement at UCSC* (Unpublished senior thesis). University of California, Santa Cruz.

Pearson, A. R., Schuldt, J. P., Romero-Canyas, R., Ballew, M. T., & Larson-Konar, D. (2018). Diverse segments of the US public underestimate the environmental concerns of minority and low-income Americans. *Proceedings of the National Academy of Sciences, 115*(49), 12429–12434.

Valencia, R. R. (2010). Dismantling contemporary deficit thinking: Educational thought and practice. Routledge.

Zelenzy, L. C., Chua, P., & Aldrich, C. (2000). Elaborating on gender differences in environmentalism. *Journal of Social Issues, 56*(3), 443–457.

Flora Lu is a Professor of Environmental Studies at the University of California, Santa Cruz and an ecological anthropologist specializing in conservation politics, environmental justice, and political ecology. She served as the Provost of College Nine and John R. Lewis College from 2014 to 2023, during which time she helped establish the People of Color Sustainability Collective (PoCSC) and the Apprenticeship in Community Engaged Research (H)ACER. She was a Postdoctoral Fellow at Stanford University and an Assistant Professor in Ecology at the University of North Carolina at Chapel Hill before joining the UCSC faculty in 2008.

Randy Uang is a Senior Survey Research Analyst in Institutional Research, Assessment, and Policy Studies at the University of California, Santa Cruz. He has experience conducting research on public health, public security, and environmental issues. He has a Ph.D. in Government from the University of Texas at Austin.

Anna Sher is a Director of Assessment and Survey Research at the University of California, Santa Cruz. She has been involved in survey research on various issues ranging from sustainability to diversity, equity and inclusion in all aspects of student experience and strategic planning for the university. She holds a Ph.D. in Sociology from Stony Brook University.

Environmental Sustainability and Epidemiological Struggle: Student Experiences of COVID-19

6

Tashina J. Vavuris, Cassandra Wood, Flora Lu, Randy Uang, and Anna Sher

Abstract

This chapter explores undergraduates' perceptions of environmental and community health in the context of COVID-19 based on a 2022 campus-wide survey at the University of California, Santa Cruz (UCSC). We explore: (1) how UCSC students experience environmental health threats in their home communities; (2) whether disparities in environmental health exposures exist among UCSC students based on race/ethnicity and first-generation status; and (3) whether students make connections between environmental health threat exposures and the impacts of COVID-19, and, if so, how that shapes students' perceptions of sustainability. We found that over half of UCSC undergraduate respondents were impacted by environmental health threats, and that a majority of students believe that these environmental health threats exacerbated their experience of the COVID-19 pandemic, thereby perceiving environmental sustainability to be more important as a result. This research points to a need to better understand student experiences of the pandemic more integratedly and comprehensively, integrating educational, economic, environmental concerns with epidemiological ones.

Keywords

Environmental justice · Environmental sustainability · Environmental health · Community health · Higher education · Student experiences of COVID-19

Introduction

After seeing how my community struggled with the pandemic and environmental sustainability issues at the same time, it has become even more important to me to learn different ways to help protect the environment and to provide food safely & ethically. (Student of Color who answered "Much More Important" on the 2022 UCSC Environmental Sustainability Survey regarding attitudes about environmental sustainability).

I've realized through the pandemic environmental public health is a large issue and certain minority groups are more susceptible to diseases and their health is more likely to deteriorate due to these environmental hazards. (Student of Color who answered "Much More Important" on the 2022 UCSC Environmental Sustainability Survey regarding attitudes about environmental sustainability).

Although my family did not have health concerns, people in our community experienced food insecurity and housing insecurity due to lack of work.

T. J. Vavuris (✉) · F. Lu · R. Uang · A. Sher
University of California, Santa Cruz, CA, USA
e-mail: tvavuris@ucsc.edu

C. Wood
University of California, Los Angeles, CA, USA

F. Lu, E. Murai (eds.), *Critical Campus Sustainabilities*, Sustainable Development Goals Series,
https://doi.org/10.1007/978-3-031-30929-8_6

(Student of Color who selected "unsafe/insecure housing" and "food insecurity" on the 2022 UCSC Environmental Sustainability Survey as factors that made the pandemic more difficult).

Environmental Justice (EJ) scholars and advocates have long argued that the definition of the 'environment' encompasses not just "the physical and natural world" but extends to where people "live, work, play, learn" (Bullard, 2018). A more expansive notion of environment centers the ubiquitous, disproportionate and normalized environmental problems burdening low income, Black, Indigenous, and People of Color (BIPOC) communities (Agyeman et al., 2002; Bullard, 1993, 1994; Taylor, 2014; Pulido, 2000; Pellow, 2017; Sze, 2020). The term "environmental racism" describes the ways that non-whites are disproportionately exposed to pollution, toxic waste and other forms of environmental harm, reflecting the "intersection of racism, social justice, political economy and its methodological and political implications" (Pulido, 1996: 142). From the Standing Rock Sioux's opposition to the Dakota Access Pipeline (DAPL), the water crisis in Flint, MI, to the contaminated land and water issues found within California's Central Valley, examples of environmental racism abound (Sze, 2020).

Relatedly, the public health issues that arise from these forms of structural injustice are also not equally distributed (Bullard & Wright, 1993)—a point clearly illustrated by the COVID-19 pandemic. According to Marya and Patel, the COVID-19 pandemic "points to a vulnerability in global health" with the ways race and class impact health outcomes (2021: 19). We have witnessed "glaring health disparities" in which low-income communities (Cole et al., 2021) and "racial and ethnic minority groups in the USA are more likely to contract, be hospitalized with, and die from COVID-19 versus whites" (Njoku, 2021: 1). These communities face heightened exposure risks because of systemic inequities in their living, working, and environmental conditions. For example, farmworkers, deemed as "essential workers" during the pandemic, were not provided support to alleviate their financial burdens, crowded living conditions, or exposure risks (Oatman, 2020). The California Institute of Rural Studies found that "farmworkers were three times as likely to test positive for COVID-19 as workers in any other industry in Monterey County" and Latinx people were "more than three times as likely as white people to die from the virus" (Oatman, 2020: 7).

In the words of Magaly Santos, a UC Berkeley undergraduate and daughter of farmworkers from Gonzales, California, "farmworkers became even more disposable than they already were," forced to continue working despite the raging pandemic, despite the apocalyptic orange skies caused by noxious wildfire smoke, without concern for their health or proper precautions in place for their safety (Oatman, 2020: 2). Living in a former labor camp used to house *braceros*, Magaly's family and neighbors are "unsheltered in place": overcrowded, unable to quarantine if ill, without alternative housing in the case of eviction or evacuation (Oatman, 2020). Low pay, high cost of living, lack of affordable housing, immigration status, limited access to medical care, limited English proficiency, insecure employment, and anti-immigrant politics are mutually constitutive and exacerbating forces that influence disease mitigation, transmission, treatment, and recovery.

As Magaly's example illustrates, students—especially people of color, low-income students, and students of other historically marginalized identities—have been especially negatively impacted by COVID-19 (Lederer et al., 2021; Njoku, 2021; Correia et al., 2022; Burt et al., 2022). Uneven and inconsistent access to learning tools such as technological devices, internet service, and teachers' attention during remote learning exacerbated already existing achievement gaps for Black, Hispanic, and Indigenous students (Dorn et al., 2020a) and inflicted learning losses that might last a lifetime (Dorn et al., 2020b). As mentioned above, low-income communities of color are particularly hard hit by the virus, so students from these backgrounds experienced increased levels of stress over concerns about their own health and that of their loved ones (Lederer et al., 2021; Wang et al., 2020). These worries, compounded by stress related to increased academic difficulty, resulted in an alarming increase in rates of depression and anxiety among students (Wang et al., 2020). The COVID-19 pandemic illustrates the interconnections between race and class, public health and educational attainment. It also offers a lens through which to explore the relevance of environmental concerns in these relationships, taking seriously the definition of environment as the places in which we live, learn, and work.

In this chapter, we ask, how, if at all, did the experience of the pandemic shape the ways that students think about environmental sustainability? While there has been research exploring how environmental conditions, such as the level of air pollution or access to green spaces, impacted physical and mental health outcomes of COVID-19 (Berman & Ebisu, 2020; Slater et al., 2020), less studied are the ways students' hometown environmental conditions mediated their experience of COVID-19 and potentially changed their perspectives about sustainability. Because mainstream environmentalism tends to be predominantly White, could something as profoundly impactful as a worldwide pandemic bring into even starker relief the entanglements of environmental racism and epidemiological precarity? We first present methods and findings from a 2022 campus-wide undergraduate sustainability survey at the University of California, Santa Cruz (UCSC) focusing on: (1) the prevalence of environmental health threats in the hometowns of UCSC students; (2) the differences found within environmental health exposures among populations of students at UCSC based on race/ethnicity and first-generation status; and (3) the interplay between environmental health threat exposures and impacts of COVID-19; and (4) the degree to which students thought differently about environmental justice and sustainability as a result.

Methods

In 2016, 2019, and 2022, UC Santa Cruz's People of Color Sustainability Collective (PoCSC) and the Institutional Research, Assessment, and Policy Studies (IRAPS) unit implemented a campus-wide undergraduate Sustainability survey. This chapter focuses on the results of the 2022 survey, which had a response rate of 13% (n = 2273 of 17,064), because it was the only one in which COVID-19 was explicitly discussed. The survey was approved by the Institutional Review Board and sought to develop more effective co-curricular programming to increase environmental inclusion and equity. All undergraduates were invited to participate in the survey where they were asked to share their perceptions of and participation in the environmental sustainability movement on campus. Survey respondents were generally representative of the student population in terms of race and ethnicity, first generation status, transfer status, and class level. Weighted data were used for our analysis to account for slight differences in response rates and proper representation of the student population in terms of gender and ethnicity.

In addition to summarizing the overall results, we also undertook analyses comparing race/ethnicity (i.e., People of Color (PoC) versus white, non-Hispanic (WNH) students), and first-generation (first-gen) students versus continuing-generation (cont-gen) students. The PoC category includes Asian American; Hispanic/Latinx; African American, Black and Caribbean; Native American; and Pacific Islander domestic students. At the time of the survey, the three largest racial/ethnic groups among UCSC undergraduates were Hispanic/Latinx (26%); Asian American (29%), and white, non-Hispanic (WNH, 31%). While there are other racial categories on campus, their numbers are low enough that they collectively make-up the remaining 14% of the student population. Of our total of 2273 survey respondents, approximately 60% were PoC (n = 1502); 29% WNH (n = 668); and 34% first-generation students (n = 778).

Drawing on a combination of Likert, closed-ended and open-ended prompts, we focus on the following questions from the survey:

1. Which environmental health threats impact your home community? Select all that apply (air contamination, water contamination, soil contamination, food contamination, food insecurity, toxic/hazardous waste exposure,

lack of greenspace, unsafe/insecure housing, inadequate sanitation, none of the above)?
2. Do you or does anyone in your immediate family suffer from environmentally caused health problems? (ex: asthma, respiratory problems, lead poisoning, etc.)?
3. Have any of these factors contributed to making the experience of the pandemic more difficult for your home community?
4. Please provide a brief example about how any of these environmental health concerns have impacted your home community or family experience during the COVID-19 pandemic.
5. Has the COVID-19 pandemic influenced your attitudes about environmental sustainability in any way?
6. Please explain your answer above: has the COVID-19 pandemic influenced your attitudes about environmental sustainability in any way?

We used a chi-square test for dichotomous categorical variables to determine if the differences were statistically significant at $p < 0.05$. For Question 2, the research team grouped responses into the following categories: student, someone in the student's family, both student and family member, and no one. For Question 3, respondents were asked to choose whether each of the environmental health threats listed made the pandemic "much more difficult," "somewhat more difficult," or "had no impact". The categories of "much more difficult" and "somewhat more difficult" were grouped/recorded into a binary/dichotomous variable. For Question 5, respondents were asked to describe their attitudinal changes using the following scale: "much more important," "more important," "not changed," "less important," and "much less important".

Findings

Community Environmental Health Threat Exposure

Across all survey respondents, the top environmental health threats impacting home communities (Fig. 6.1) are air contamination (55%), unsafe/insecure housing (44%), lack of greenspace (37%), food insecurity (37%), and water contamination (33%). Given that multiple threats can co-exist, the percentages do not add up to 100%. Note that only 20% of the respondents said that their home communities did not face any of these environmental health threats.

Figure 6.2 shows the two categories that were statistically significant when we undertook the comparison by race/ethnicity. Food insecurity and contamination are problems more commonly experienced by PoC students' home communities compared to those of WNH students.

Figure 6.3 shows that compared to continuing-generation students, first-generation students' home communities more often experience a variety of environmental concerns, namely unsafe/insecure housing, food insecurity, lack of greenspace and toxic-hazardous waste exposure.

In response to Question 2, which asks if the student or anyone in their immediate family suffers from environmentally caused health problems, we found that 16% of respondents reported that they themselves are suffering from environmentally caused health problems. One-third (33%) of respondents reported that someone else in their family suffers from environmentally caused health problems. These two situations can overlap, and in fact, 11% of respondents reported that both themselves and at least one other person in their family suffer from environmentally caused health problems.

We further analyzed this question by race/ethnicity and first-generation status, and did not find any statistically significant differences except for one. For PoC respondents, 17% (vs. 16% WNH) of students suffer from environmentally caused health threats; in 35% (vs. 31% WNH) of cases, someone in the student's family suffers from environmentally caused health problems; and 11% (same for WNH students) of the time, both the student and someone in student's family suffer from environmentally caused health problems. For First-gen respondents, 17% (vs. 16% cont-gen) of students suffer from environmentally caused health threats; in 40% (vs. 30% cont-gen) of cases, someone in the student's family suffers from environmentally caused health problems; and 12% (vs. 10% cont-gen) of the

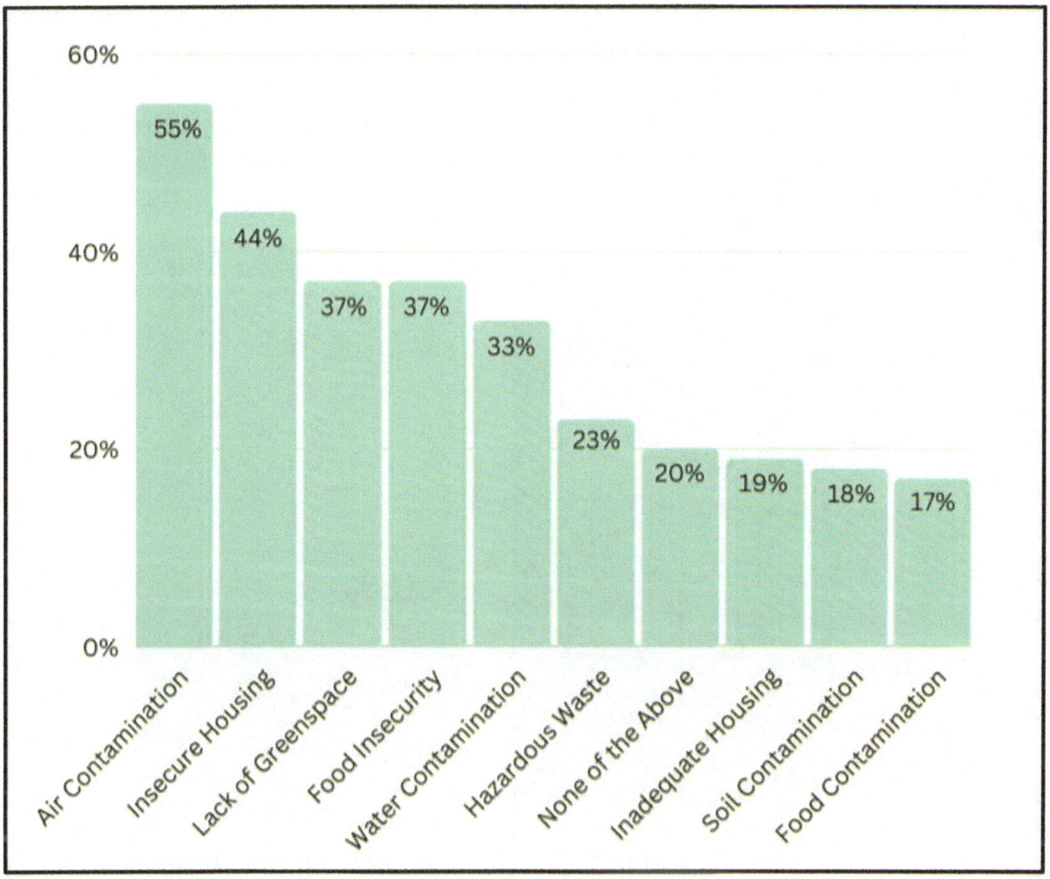

Fig. 6.1 Frequency of reported environmental health threats by UCSC students in 2022

time both student and someone in student's family suffer from environmentally caused health problems. The finding that 40% of first-generation students report a family member suffering from environmentally caused health issues, compared to 30% of continuing-generation students, is statistically significant.

COVID-19 + Environmental Health Threats

All survey respondents reported that environmental health threats made their COVID-19 experience either "much more difficult" or "somewhat more difficult" (Fig. 6.4). The top issues, reported by over 60% of respondents, were (in descending order): food insecurity, unsafe/ insecure housing, inadequate sanitation, lack of greenspace, air contamination, and food contamination.

Figure 6.5 shows how various environmental health threats were experienced at a higher rate by PoC compared to WNH students: food insecurity (86% PoC vs. 80% WNH); inadequate sanitation (84% PoC vs. 67% WNH); water contamination (51% PoC vs. 32% WNH); and soil contamination (42% PoC vs. 22% WNH).

Between first-generation students and continuing-generation students, we found almost double the number of statistically significant differences (Fig. 6.6). As compared to continuing-generation students, more first-generation undergraduate students found that the environmental health threats made their COVID-19 experience worse ("much more difficult" or

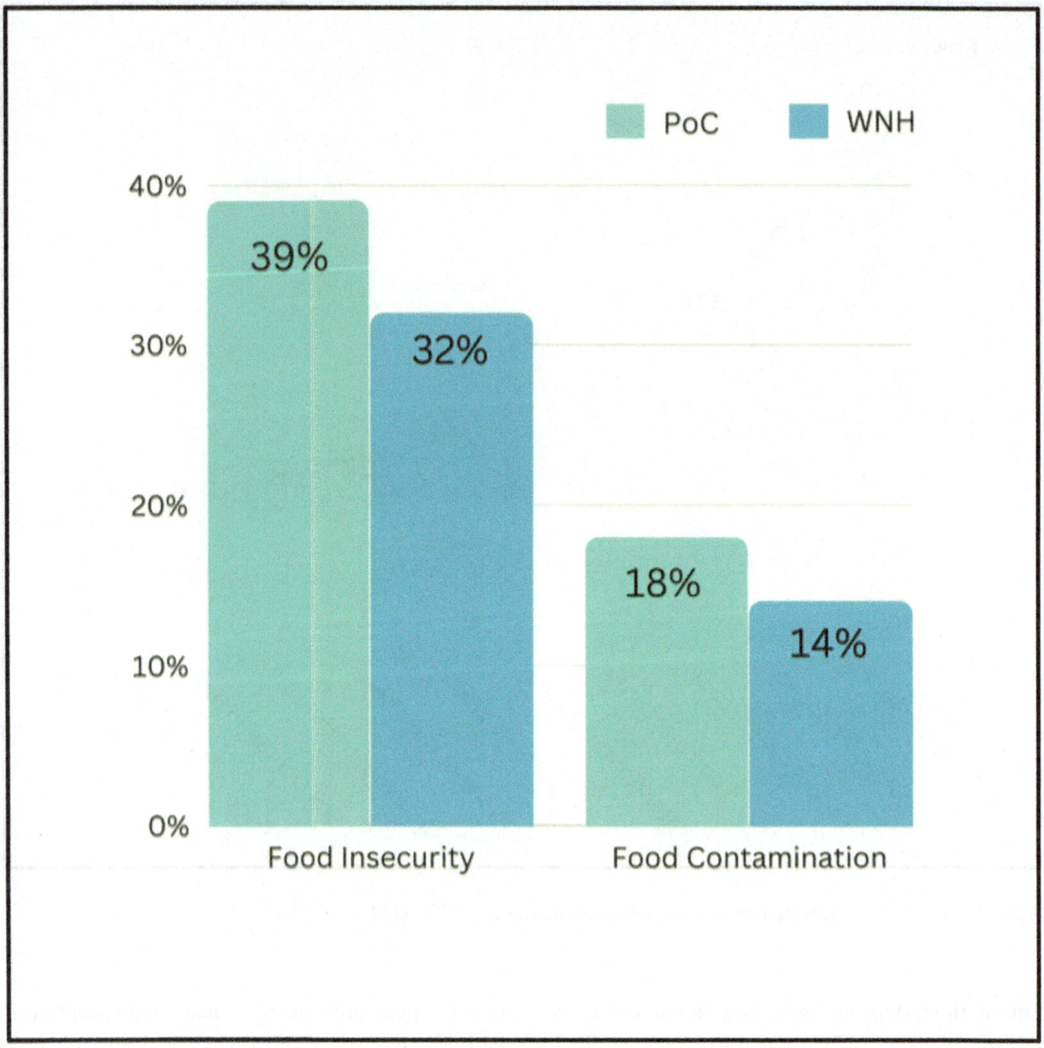

Fig. 6.2 Comparison of statistically significant environmental health threats by PoC vs. WNH in 2022

"somewhat more difficult"): food insecurity (89% first-gen vs. 82% cont-gen); inadequate sanitation (89% first-gen vs. 73% cont-gen); unsafe/insecure housing (87% first-gen vs. 78% cont-gen); food contamination (77% first-gen vs. 54% cont-gen); air contamination (72% first-gen vs. 60% cont-gen); toxic/hazardous waste exposure (58% first-gen vs. 43% cont-gen); and soil contamination (51% first-gen vs. 33% cont-gen).

Paired with Question 3, an open-ended answer explanation section (Question 4) was available if students wanted to elaborate on their answers. Undergraduates were asked to "provide a brief example about how any of these environmental health concerns have impacted your home community or family experience during the COVID-19 pandemic." These responses provided important context. For example, events such as the Northern California wildfires exacerbated existing disparities in exposure to environmental health threats. Additionally, students shared their experiences around food and housing insecurity:

> Although my family did not have health concerns, people in our community experienced food insecurity and housing insecurity due to lack of work.

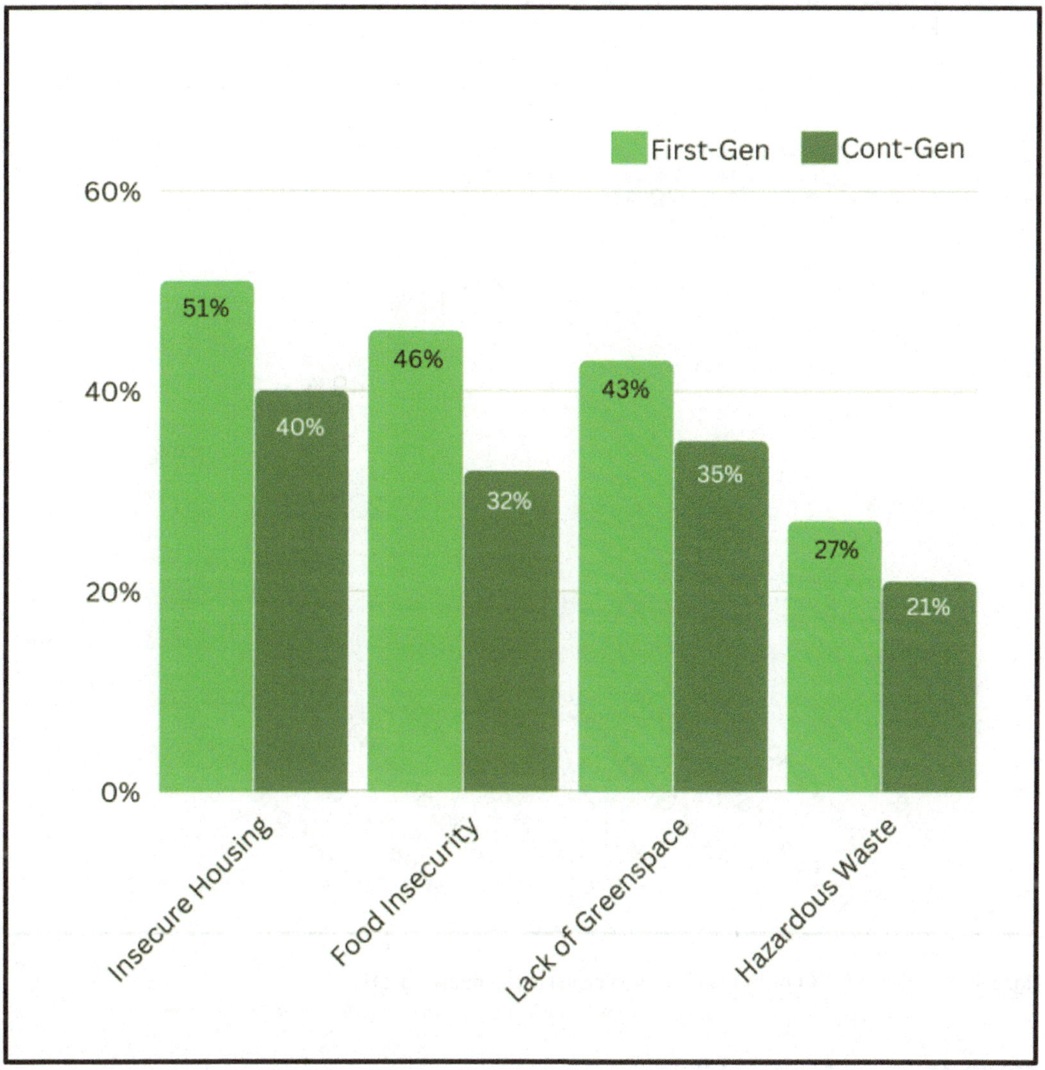

Fig. 6.3 Comparison of statistically significant environmental health threats by First-generation vs. Continuing-generation students, in 2022

(Student of Color who selected "unsafe/insecure housing" and "food insecurity" on the 2022 UCSC Environmental Sustainability Survey as factors that made the pandemic more difficult)

My uncle died because of a combination of unsafe/insecure housing and inadequate access to healthy food during this pandemic. Additionally, many of the people I know who are disabled for various reasons have had to ration their medication at times. (Student of Color who selected "air contamination", "water contamination", "lack of greenspace", "unsafe/insecure housing", and "inadequate sanitation" on the 2022 UCSC Environmental Sustainability Survey as factors that made the pandemic more difficult)

Twice during the pandemic while at UCSC I have had to choose between finding affordable housing and food or going to class. (White, non-Hispanic student who selected "unsafe/insecure housing" and "food insecurity" on the 2022 UCSC Environmental Sustainability Survey as factors that made the pandemic more difficult)

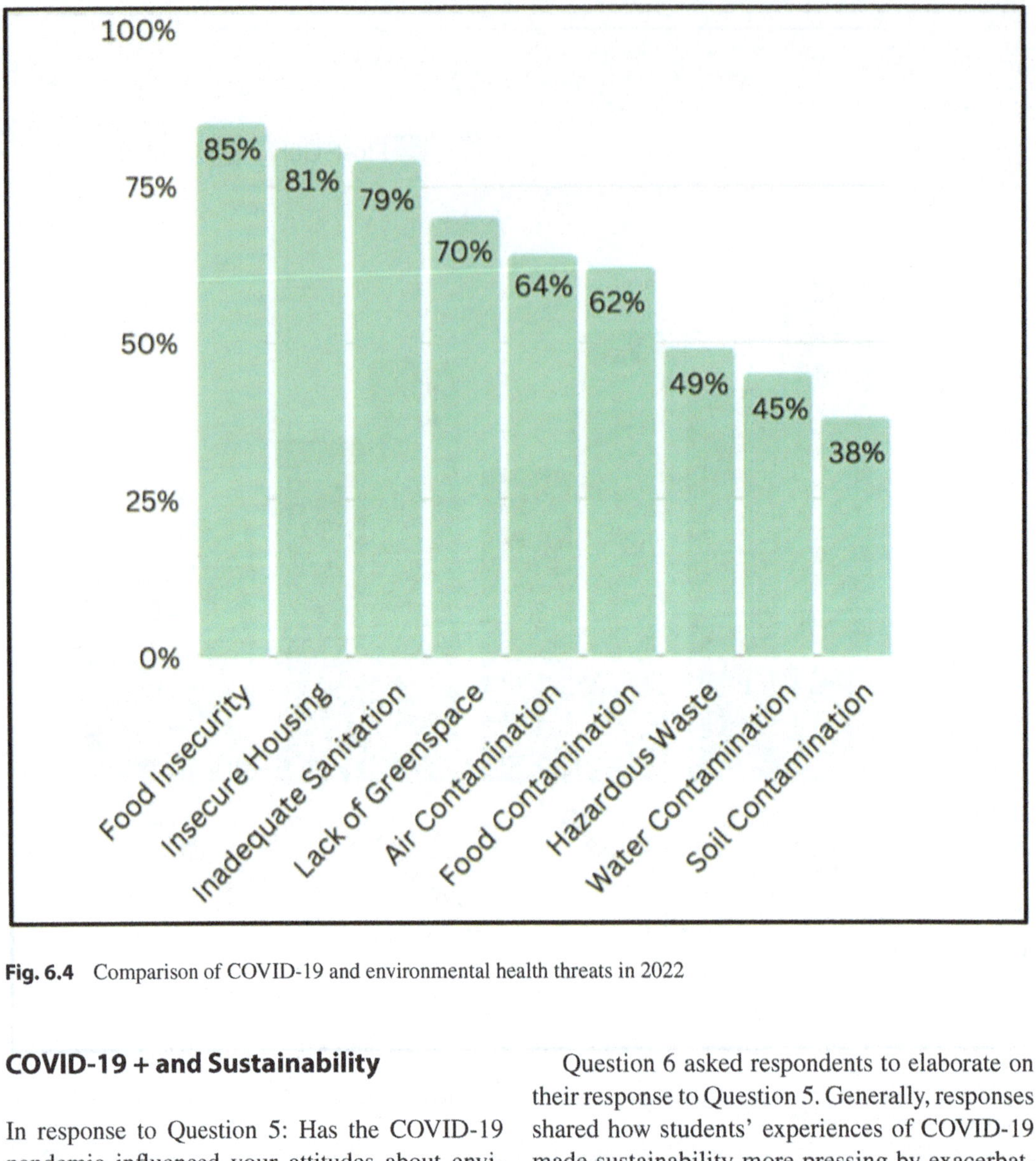

Fig. 6.4 Comparison of COVID-19 and environmental health threats in 2022

COVID-19 + and Sustainability

In response to Question 5: Has the COVID-19 pandemic influenced your attitudes about environmental sustainability in any way?, we found that 61% of respondents indicated that environmental sustainability became "more important" or "much more important". We found that students from historically marginalized groups especially made connections between public health and sustainability: more students of color (64%) than WNH (55%) students, and more first-generation students (68%) than continuing-generation students (58%) said that environmental sustainability had become "more important" or "much more important".

Question 6 asked respondents to elaborate on their response to Question 5. Generally, responses shared how students' experiences of COVID-19 made sustainability more pressing by exacerbating already difficult living conditions. The responses reveal an evolving understanding of the ways the pandemic affected housing and food security, which students see as environmental issues. As an example, a student of color responded by stating, "extremely variable weather and increased environmental disaster will probably influence people in much the same way as the pandemic (via food and housing insecurity), so environmental sustainability has definitely become a more pressing issue." This response points to how environmental problems—

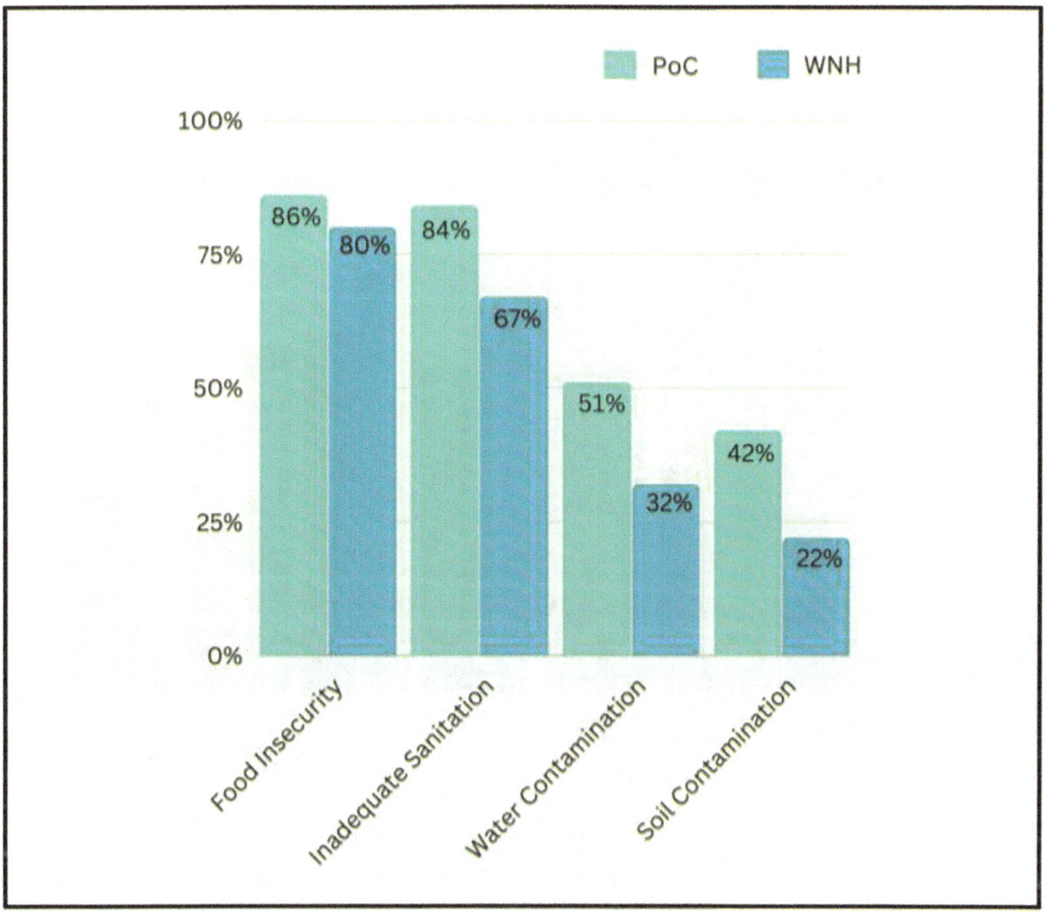

Fig. 6.5 Comparison of statistically significant environmental health and COVID-19 threats by PoC versus WNH in 2022

such as impacts from climate change—will influence how people experience environmental health threats (e.g., air contamination, insecure housing, lack of greenspace, food insecurity, water contamination). Conversely, there were some students for whom sustainability decreased in importance after the onset of COVID-19. For example, one respondent, a student of color who marked "less important" regarding the importance of sustainability during COVID-19, stated that "the way COVID has impacted many different industry operations it feels near impossible to worry about the environmental impact when people are dying." This student, and others like them, do not perceive their wellbeing, immediate concerns and basic needs as environmental or sustainability related, and do not make the connection that the pandemic is fundamentally and inextricably ecological, social, economic and political. Education in critical sustainabilities must prioritize challenging such compartmentalized thinking, and must overcome entrenched notions that environmental concerns are separate, non-essential or less urgent.

Discussion

This research illuminates the interconnectedness of environmental sustainability, education, and epidemiology. First, we found that students experience numerous environmental health threat

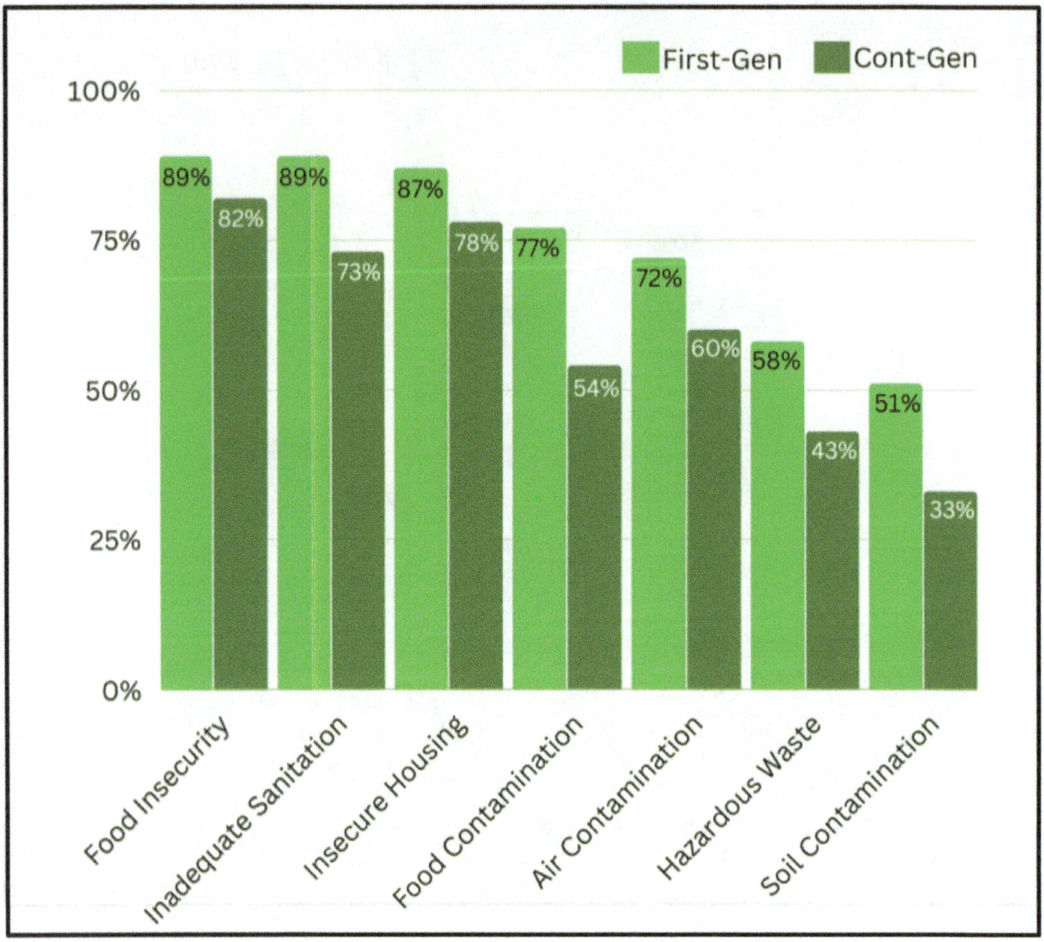

Fig. 6.6 Comparison of statistically significant environmental health threats and COVID-19 by First-generation versus Continuing-generation in 2022

exposures (e.g., air pollution, water contamination, food insecurity, insecure housing) and environmentally caused health problems (e.g., asthma, respiratory problems, lead poisoning). Over half of the undergraduate respondents at UCSC indicated that environmental health threats impact their home communities. Second, students of color experience certain exposures, such as food contamination and food insecurity, at statistically significantly higher rates compared with White students. First-generation students are more likely to be impacted by food insecurity, toxic/hazardous waste exposure, lack of greenspace, and unsafe/insecure housing as compared continuing-generation students, who come from homes where at least one of their parents attended a four-year college or university.[1] Third, this myriad of environmental inequities exacerbated the COVID-19 experiences of UCSC students, their families, and communities.

We found that students (and their families) are not only experiencing environmental health threats and health problems, but that their COVID-19 experiences were exacerbated as a result. For over 60% of the student body as a whole, and for two-thirds of students of color and

[1] Being a first-generation student is not reducible to or synonymous with low socio-economic status, but they are related, as higher education is an important means of upward economic mobility. Even when socio-economic status increases, the first-generation disadvantage is not eliminated (Wilbur & Roscigno, 2016).

first-generation students, environmental sustainability became more important as a result of the pandemic.

The vulnerabilities faced by communities experiencing environmental injustice were laid bare during the COVID-19 pandemic. For example, there are higher mortality rates from COVID-19 in areas with higher air pollution levels, which are found, due to systemic racism, in BIPOC communities (Soares, 2021). Long-term exposure to air pollution—especially fine particulate matter known as $PM_{2.5}$, which comes from fuel combustion (i.e., vehicles, refineries, power plants)—increases people's vulnerability to experiencing the most severe COVID-19 outcomes (Friedman, 2020). Tessum et al. (2019) documented "pollution inequity", defined as:

> …the difference between the environmental health damage caused by a racial–ethnic group and the damage that group experiences. We show that, in the United States, $PM_{2.5}$ exposure is disproportionately caused by consumption of goods and services mainly by the non-Hispanic white majority, but disproportionately inhaled by black and Hispanic minorities. On average, non-Hispanic whites experience a "pollution advantage": They experience ~17% less air pollution exposure than is caused by their consumption. Blacks and Hispanics on average bear a "pollution burden" of 56% and 63% excess exposure, respectively, relative to the exposure caused by their consumption (6001).

This environmental/epidemiological nexus manifests in the disproportionate toll that the pandemic has taken on people of color. As compared to non-Hispanic whites, Black people have 2.1 times the hospitalization rate; Hispanic/Latinx people have 1.9 times the hospitalization rate; and Indigenous people have 2.5 times the hospitalization rate and 2.1 times the death rate (CDC, 2022).

For approximately 40% of the students we surveyed, the pandemic did not increase the importance they ascribe to environmental issues. For many of these students, it is because they already are extremely concerned about sustainability. However, there are those who articulate that environmental concerns must take a back seat when "people are dying" from the virus, an erroneous dichotomy between public health and environmental sustainability that calls for environmental education to do a better job at conveying how sustainability is inextricable from confronting systemic injustices. Our research highlights that such educational efforts should recognize the conjoined public health and environmental harms prevalent among so many of our undergraduates, especially PoC and first-generation, and build more curricula that reflect their lived experiences and values.

Conclusion

Environmental problems and exposure to environmental risks impact students, their home communities, and the health of both. Though COVID-19 was a "once-in-a-lifetime" pandemic (Guterres, 2020), it was not an isolated incident in the way it impacted students' day-to-day lives. With environmental precarity and environmental concerns on the rise, it is important to think about the vital interconnectedness of education, environment, and epidemiology and the way students are impacted by multitude of challenges that arise out of this triangulation. We echo the demand "that the fight for environmental justice be at the center of COVID-19 recovery" (Njoku, 2021: 6). UCSC undergraduate students are experiencing exposure to environmental health threats ranging from air contamination to food insecurity that mirror growing environmental precarity worldwide (Igini, 2022). In this chapter, we suggest the importance of connecting social and ecological aspects of environmental framings to better understand environmental problems and how different populations experience those problems. As Iles and Montenegro de Wit (2020: 659) remind us, times of crisis render even more stark the "hierarchies of knowledge and malignant power inequalities" among decision makers, experts, and scientists whose determinations "make a tremendous difference to the health and socioeconomic impacts that people of color, rural communities, farmers, workers, and consumers experience in times of catastrophe." To this end, the threads of this conversation connect to other chapters within this edited volume that speak to the importance of uprooting the inequities, leaning into the complexities, centering diverse epistemological and cultural approaches, and embracing relational teaching practices and critical research methodologies to understand the interconnectedness of social and environmental awareness.

References

Agyeman, J., Bullard, R. D., & Evans, B. (2002). Exploring the nexus: Bringing together sustainability, environmental justice and equity. *Space and polity, 6*(1), 77–90. https://doi.org/10.1080/13562570220137907

Berman, J. D., & Ebisu, K. (2020). Changes in US air pollution during the COVID-19 pandemic. *Science of the Total Environment, 739*.

Bullard, R. D. (1993). The threat of environmental racism. *Natural Resources & Environment, 7*(3), 23–56.

Bullard, R. D. (1994). Environmental justice for all: It's the right thing to do. *Journal of Environmental Law & Litigation, 9*, 281.

Bullard, R. D. (2018). *Dumping in Dixie: Race, class, and environmental quality*. Routledge.

Bullard, R. D., & Wright, B. H. (1993). Environmental justice for all: Community perspectives on health and research. *Toxicology and Industrial Health, 9*(5), 821–841.

Burt, K. G., Eubank, J. M., & Orazem, J. (2022). Female black, indigenous, and students of color demonstrate greater resilience than other students during a global pandemic. *Equity in Education & Society*, 27526461221105094.

Centers for Disease Control and Prevention (CDC). (2022). *Risk for COVID-19 infection, hospitalization, and death by race/ethnicity*. Retrieved January 24, 2023, from https://www.cdc.gov/coronavirus/2019-ncov/covid-data/investigations-discovery/hospitalization-death-by-race-ethnicity.html

Cole, H. V. S., Anguelovski, I., Baró, F., García-Lamarca, M., Kotsila, P., Pérez del Pulgar, C., & Triguero-Mas, M. (2021). The COVID-19 pandemic: Power and privilege, gentrification, and urban environmental justice in the global north. *Cities & Health, 5*(sup1), S71–S75. https://doi.org/10.1080/23748834.2020.1785176

Correia, K. M., Bierma, S. R., Houston, S. D., Nelson, M. T., Pannu, K. S., Tirman, C. M., Cannon, R. L., Clance, L. R., Canterbury, D. N., Google, A. N., Morrison, B. H., & Henning, J. A. (2022). Education racial and gender disparities in COVID-19 worry, stress, and food insecurities across undergraduate biology students at a southeastern university. *Journal of Microbiology & Biology Education, 23*(1). https://doi.org/10.1128/jmbe.00224-21

Dorn, E., Hancock, B., Sarakatsannis, J., & Viruleg, E. (2020a). COVID-19 and learning loss—Disparities grow and students need help. *McKinsey & Company, 8*, 6–7.

Dorn, E., Hancock, B., Sarakatsannis, J., & Viruleg, E. (2020b). COVID-19 and student learning in the United States: The hurt could last a lifetime. *McKinsey & Company, 1*.

Igini, M. (2022). *Top 6 Environmental Issues in the US in 2022*. Retrieved August 6, 2022, from https://earth.org/top-environmental-issues-us/

Friedman, L. (2020, April 7). New research links air pollution to higher Coronavirus death rates. *The New York Times*. https://www.nytimes.com/2020/04/07/climate/air-pollution-coronavirus-covid.html

Guterres, A. (2020). All hands on deck to fight a once-in-a-lifetime pandemic. *United Nations COVID-19 response*. Retrieved January 24, 2023, from https://www.un.org/en/un-coronavirus-communications-team/all-hands-deck-fight-once-lifetime-pandemic

Iles, A., & Montenegro de Wit, M. (2020). Who gets to define 'the COVID-19 problem'? Expert politics in a pandemic. *Agriculture and Human Values, 37*(3), 659–660.

Lederer, A. M., Hoban, M. T., Lipson, S. K., Zhou, S., & Eisenberg, D. (2021). More than inconvenienced: The unique needs of US college students during the COVID-19 pandemic. *Health Education & Behavior, 48*(1), 14–19.

Njoku, A. U. (2021). COVID-19 and environmental racism: Challenges and recommendations. *European Journal of Environmental Public Health, 5*.

Marya, R., & Patel, R. (2021). *Inflamed: Deep medicine and the anatomy of injustice*. Penguin.

Oatman, M. (2020, October 2). *Unsheltered in place*. Mother Jones. https://www.motherjones.com/environment/2020/10/salinas-housing-coronavirus-farmworkers-wildfires-lettuce-salad-crops/

Pellow, D. N. (2017). *What is critical environmental justice?* Wiley.

Pulido, L. (1996). A critical review of the methodology of environmental racism research. *Antipode, 28*(2), 142–159.

Pulido, L. (2000). Rethinking environmental racism: White privilege and urban development in Southern California. *Annals of the Association of American Geographers, 90*(1), 12–40. https://doi.org/10.1111/0004-5608.00182

Slater, S. J., Christiana, R. W., & Gustat, J. (2020). Recommendations for keeping parks and green space accessible for mental and physical health during COVID-19 and other pandemics. *Preventing Chronic Disease, 17*.

Soares, K. (2021). An inequitable pandemic. How environmental racism has worsened COVID-19 in communities of color. *ClimateXChange*. https://climate-xchange.org/2021/09/23/an-inequitable-pandemic-how-environmental-racism-has-worsened-covid-19-in-communities-of-color/

Sze, J. (2020). *Environmental justice in a moment of danger*. University of California Press.

Taylor, D. (2014). *Toxic communities: Environmental racism, industrial pollution, and residential mobility*. New York University Press.

Tessum, C. W., Apte, J. S., Goodland, A. L., & Hill, J. D. (2019). Inequity in consumption of goods and services adds to racial-ethnic disparities in air pollution exposure. *Proceedings of the National Academy of Sciences, 116*(13), 6001–6006. https://doi.org/10.1073/pnas.1818859116

Wang, X., Hegde, S., Son, C., Keller, B., Smith, A., & Sasangohar, F. (2020). Investigating mental health of US college students during the COVID-19 pandemic: Cross-sectional survey study. *Journal of Medical Internet Research, 22*(9).

Wilbur, T. G., & Roscigno, V. J. (2016). First-generation disadvantage and college enrollment/completion. *Socius, 2*. https://doi.org/10.1177/2378023116664351

Tashina J. Vavuris is a Ph.D. candidate in the Environmental Studies Department at the University of California, Santa Cruz. She received her Master's degree in Sociology from San Diego State University, where she examined critical pedagogies and community-based practices. Her current research interests relate to environmental justice education with the intersection of critical pedagogies, participatory approaches, and Indigenous science relating to environmental sense-making and praxis.

Cassandra Wood is a graduate from University of California, Santa Cruz in Environmental Studies and is currently in the Master's degree program in the Urban and Regional Planning department at University of California, Los Angeles. Their research lies in socio-spatial injustice in the built environment to perpetuate homogenous enclaves via spatial accessibility. Aside from urban planning, their interests are gastro-imperialism and colonization.

Flora Lu is a Professor of Environmental Studies at the University of California, Santa Cruz and an ecological anthropologist specializing in conservation politics, environmental justice, political ecology and Indigenous resource governance. She served as the Provost of College Nine and John R. Lewis College from 2014 to 2023, during which time she helped establish the People of Color Sustainability Collective (PoCSC) and the Apprenticeship in Community Engaged Research (H)ACER. She was a Postdoctoral Fellow at Stanford University and an Assistant Professor in Ecology at the University of North Carolina at Chapel Hill before joining the UCSC faculty in 2008.

Randy Uang is a Senior Survey Research Analyst in Institutional Research, Assessment, and Policy Studies at the University of California, Santa Cruz. He has experience conducting research on public health, public security, and environmental issues. He has a Ph.D. in Government from the University of Texas at Austin.

Anna Sher is a director of Assessment and Survey Research in Institutional Research, Assessment, and Policy Studies at the University of California, Santa Cruz. She has been involved in survey research on various issues ranging from sustainability to diversity, equity and inclusion in all aspects of student experience and strategic planning for the university. She holds a Ph.D. in Sociology from Stony Brook University.

Critical Environmentalisms: Overcoming Institutional Obstacles to Meet Students' Demands for Sustainability Curricula and Action

7

Flora Lu, Emily Murai, Serena Campbell, and Hillary Angelo

Abstract

This chapter brings together the voices of undergraduates, faculty and staff to share their perspectives about whether UC Santa Cruz is doing enough to address environmental concerns, and how it can better support critical sustainability and action. In the first section, we report the longitudinal results of surveys conducted in 2016, 2019 and 2022 of over 8000 respondents in total. Students' self-reported environmental concern increases over time, and is mostly similar for white and non-white undergraduates, except for a few environmental issues. However, the disparity between students' environmental concern and how they perceive the campus to be prioritizing environmental issues grows over time, with students perceiving the university as falling short on addressing all environmental issues surveyed. We then shift focus to environmental curricula at UC Santa Cruz, exploring the results from the 2022 data about coverage of a range of environmental topics, how much students reported learning about environmental concerns, and what they wanted to see emphasized in their coursework. One of the major findings is that students of color, by a large margin, want environmental curricula to have greater racial, ethnic, gender, etc. representation. To this end, the chapter shares results of a new project, Critical Environmentalisms, in which faculty and staff were asked how the institution could support environmental efforts and education that is connected to issues of power, difference, identity and equity.

Keywords

Climate justice · Youth organizing · Curricular change · Institutional shortcomings · Faculty · Surveys

Introduction

On November 19, 2022, Maya Caminada stood in front of a group of around 40–50 attendees who had converged on a Saturday morning at Rachel Carson College. The People of Color Sustainability Collective intern and chief event coordinator of this first annual Student Climate Conference at the University of California, Santa Cruz (UCSC) spoke about the pervasive sense of fear, anger and hopelessness among her

F. Lu (✉) · E. Murai · S. Campbell · H. Angelo
University of California, Santa Cruz, CA, USA
e-mail: floralu@ucsc.edu

F. Lu, E. Murai (eds.), *Critical Campus Sustainabilities*, Sustainable Development Goals Series,
https://doi.org/10.1007/978-3-031-30929-8_7

generation witnessing climate disruption and environmental degradation all around them. “We are beyond this” failure to act, she declared. “We are beyond being lied to. That is why we are calling this conference ‘Beyond the Bullsh*t: Waking Up a Warming World’.” She handed the microphone to Becca Mansergh, who read a short story about how oil seeps into her dreams, imbuing her with dread and guilt, leaving a residue like blood at a crime scene. Her words left no doubt as to the degree of eco-anxiety and eco-grief felt among these undergraduates.

The keynote speaker for the conference was Tamarah Minami, a student at Santa Cruz High School and an activist with Youth for Climate Justice. The choice was strategic, one meant to challenge notions of expertise on a college campus, and to position the commitments and priorities of Gen Z. Tamarah’s remarks, while brief, did not mince words. She connected climate change and gun violence; demanded more environmental curricula in the schools; and spoke about her organization’s campaigns, from coordinating walkouts at schools to passing legislation that would provide free bus passes for people under age 25 in CA, an effort vetoed by Governor Gavin Newsom. “Old, white politicians and the 1% will not do anything that is not beneficial to them,” Tamarah said. She stated that issues of accessibility and affordability of mass transit reflect a commitment to addressing racism and classism. Despite her notable efforts to date, Tamarah does “not want to be a climate activist as my job.” She cares about the earth and takes action because she cares, but clearly should not have to hold the burdens of combatting existential environmental threats on her slim shoulders. Members of this generation have not been given much choice.

While many of the attendees were Environmental Studies majors, the conference appealed to a cross section of the student body. One such example was Alexandra Mendoza, an Applied Math major who shared that she did not have any coursework at UCSC around themes of environmental justice, except that her Writing 2 course touched on environmental topics. She was at the conference because “climate justice is always in the back of my mind.” Her hometown of Fontana, CA is the location of a massive Amazon Distribution Center located north of her high school. Plans are in the works for another warehouse south, sandwiching the school. Her neighborhood is predominantly Mexican, Filipino and Black, and the truck traffic means that they are “sitting in pollution,” resulting in asthma and lung cancer. Families in her community and friends of hers were displaced with the arrival of the retailer. “In the nicer, white part of town, there are no warehouses,” Alexandra said. In her Writing 2 class, they read a paper about a landfill being located in a Black community; the similarities with her hometown were glaring, and talking about these issues in her class made her “feel seen.” She wants more courses to talk about issues of environmental justice—more students writing papers, giving presentations, and sharing their first-hand knowledge. Alexandra said, “Faculty should support students, uplift our voices, and make sure that the administration hears us and what we want.”

In this chapter, we explore how perceptions of environmental issues have changed over the last half a dozen years, along with the larger political climate and student demographics at UCSC. We do this by sharing some longitudinal survey data to highlight the overwhelming environmental concern evinced by our students, and their opinions about how much the campus prioritizes environmental sustainability in comparison. Additionally, we report on the most recent results of survey questions about environmental curricula at UCSC: what is being learned, how effective sustainability education is, and what students want more of moving forward. Some of the responses show statistically significant differences between students who identify as a person of color (PoC) and those who identify as non-Hispanic white (WNH), suggesting that racial differences shape environmental priorities and perceptions, and that the campus should incorporate more approaches that appeal to its diverse student body as a means to foster a greater sense of belonging. Of course, curricular decisions are made by faculty, and we then shift the discussion to include the voices of UCSC faculty who are interested in providing an environmental education that centers on themes of power relations, (in)equity, accessibility, epistemology, and social (in)justice. These faculty, some of whom experience marginalization in their departments for their approaches, share ideas about the institutional changes necessary to provide a more critical environmental education that students are demanding.

UCSC, Then and Now

UCSC is one of the youngest campuses in the UC system, and arguably the most picturesque. In 1961, the UC Regents established another UC campus on Cowell Ranch, 200 acres of coastal meadow and forest, choosing to construct a "city on the hill" just outside the main center of the seaside town of Santa Cruz over establishing a campus in the urban center of nearby San Jose (Reti et al., 2020a; Townsend, 2015). The earliest founders envisioned creating a small liberal arts college system within the confines of a major university as an alternative to the large class sizes and impersonal teaching they observed taking place at UC's flagship institutions. In the first decade of UCSC's existence, they created smaller, themed residential colleges that housed faculty alongside students in dormitories, tried to dissolve the boundaries of traditional disciplines by emphasizing interdisciplinary teaching and learning, created a personalized narrative evaluation system instead of grades, and encouraged experiential teaching like the use of encounter groups popularized in the 1970's in the classroom. The natural beauty of the campus—nestled on a remote hillside that offers sweeping views of the Monterey Bay on one side and a redwood forest on the other—became a source of inspiration for many and a primary resource for learning. The campus recruited faculty from nearby urban campuses like San Francisco State University and UC Berkeley, who brought their direct involvement with Civil Rights, anti-war and the Ethnic Studies movements to inspire the next generation of campus activists. And in the first couple of decades, the campus, leveraging its location at the center of coastal farming communities, developed campus gardens and a farm that became major sites for experiential farming, for which UCSC maintains a world renown reputation today (Reti et al., 2020a).

Much like the counterculture movement of the 1970's, the students who populated UCSC's early decades were white and of high socio-economic status, attracted to the experimental education movement and activist undertones. Over the almost six decades since its founding, the campus has both built off its reputation for experimental environmental education and progressive political climate while enlarging the demographic profile of its student body, though historically, UCSC has been one of the least racially and socio-economically diverse campuses of all the UC's (Reti et al., 2020a: 221–224; University of California Accountability Report, 2015; University of California Information Center, 2022a, b). In their two-volume compilation of oral histories spanning UCSC's first 50 years, Reti et al. (2020a, b) highlight the challenges faced by the handful of students of color that populated the campus in the early years. One interviewee who attended UCSC in the 1970s poignantly recalls,

> I grew up in neighborhoods with heavy concentrations of ethnic populations. I went to East LA. College. East L.A. College is about 50 percent Chicano, about 10 percent Negro, about 10 percent Asian-American, and the rest would be white. When I came up here, it was a total reversal, because the campus was entirely white. At that time, there were about two black students in Stevenson, maybe about two or three Chicanos, and a few Asian Americans. When I was first walking on campus, I had some kind of feeling of unhappiness. I couldn't really figure out why. As I look back on it now, I realize that this was the reason. (Reti et al., 2020a: 219)

The campus increased its population of students of color throughout the 1980s and 1990s as the demographics of the state of California shifted (Reti et al., 2020b). The student of color population has grown such that today, UCSC proudly proclaims its current status as a Hispanic-Serving Institution (HSI) and Asian American and Native American Pacific Islander-Serving Institution

(AANAPISI), designations that signify that the population of each group makes up more than 25% of the overall student population. Despite this, the UCs, across all campuses, still fall short in reflecting the demographics of the state. UCSC in particular has one of the highest campus percentages of students who identify as white (~30%), while Black and Indigenous/Native students remain overwhelmingly underrepresented (University of California Accountability Report, 2022), their growth likely hindered by voter-enacted Proposition 209 in the 1990s that prohibited the use of race as a factor in admissions (Bleemer, 2022). At the same time, in recent years, the campus has been ranked highly on indices of social mobility, as it serves a large number of Pell Grant holders (an indicator of low-income background) (Hernandez-Jason, 2020), thus providing a pathway into the middle-class, and acting as a key player in California's "upward mobility machine" (Leonhardt, 2015).

The fact that the campus is now a leader in both certain forms of environmentalism and educates wide swaths of diverse student populations can be reflected in some of the tensions raised in earlier chapters of the book and in the chapter's previous section that suggest that the campus's approach to environmentalism has not caught up with interests and the needs of the more diverse student body. Feelings of isolation and marginality among students of color persist, particularly within environmental disciplines that are disproportionately made up of students who identify as white (Dare, 2021). We provide concrete suggestions to minimize the gap between institutional approaches and student concerns around environmentalism at the end of the chapter.

Trends in Undergraduate Environmental Concern

The People of Color Sustainability Collective (PoCSC) administered surveys in 2016, 2019 and 2022 to understand student perspectives on sustainability and environmental aspects of campus. All enrolled students were invited to take an online survey where they were asked to share their perceptions of and participation in the environmental sustainability movement on campus. The survey response rates were 21% (n = 3266) in 2016; 15% (n = 2546) in 2019; and 13% (n = 2273) in 2022. Survey respondents were representative of the student population in terms of race and ethnicity, first generation status, transfer status, and class level. Weighted data was used for our analysis to ensure proper representation of the student population in terms of gender and ethnicity. In addition to the survey data, the analyses incorporated institutional data on students' demographic background (gender, race/ethnicity, first generation status) and academic path (major, college, class level, and transfer status). The authors worked closely with UCSC's Institutional Research, Assessment and Policy Studies (IRAPS) office with survey design, administration, and data analysis. This research was approved by the Institutional Review Board.

The survey instrument included Likert scale questions asking respondents to rate their personal level of importance for a variety of issues (Not important, Somewhat Important, Important, Very Important). The issues are listed below. It is important to note that there were some oversights in survey design in 2016 that were corrected in 2019 and 2022. In 2016, climate change was not listed, and food justice was framed as "access to healthy, affordable and culturally appropriate food". In 2019, we named food security and food sovereignty separately and removed the food justice definition. These changes remained in the 2022 survey.

- **Environmental Health** (address chemical and toxic exposures in air, water, soil and food in natural and built environment)
- **Agroecology** (promote the productivity, stability and equitability of agroecosystems through mimicking natural processes)
- **Conservation of Natural Resources & Protection of Biodiversity and Habitats**
- **Food Security** (access to healthy and affordable food as a basic need)
- **Food Sovereignty** (food as central to cultural identity as well as nourishment, such as revitalizing native and local foodways)

- **Environmental Justice** (intersection of race, class and power in environmental policy making and distribution of burdens of environmental degradation)
- **Climate Change** (human-caused, disruptive shifts in climate patterns regionally and globally)
- **Non-Fossil Fuel, Renewable Energy**
- **Green Building Design**
- **Shared, Accessible, and Low-Emission Transportation**
- **Waste Reduction & Prevention**
- **Water Conservation**

Figure 7.1 reports students' self-reported environmental concern on a range of issues in 2016, 2019, and 2022. It shows the percentage of UCSC undergraduate respondents who said that each particular environmental issue was important or very important to them. The trend is clear: environmental concerns rise over time, with eight of the twelve issues, or two-thirds, ranked at or above 90% in 2022.

Figure 7.2 breaks down self-reported environmental concern by race, comparing people of color (PoC) and white non-Hispanic (WNH) students. PoC are grouped first, reporting the results for the three survey years, followed by WNH. A sharp increase from 2016 for PoC (dark blue) and WNH (yellow) to 2019 is visible, a trend that is largely sustained or slightly increases in 2022.

Figure 7.3 is the same as Fig. 7.1, except that instead of reporting how important the environmental issue was to them, students were asked to share how important the issue was to the UCSC campus. The question stated, "Based on what you know about UCSC, please evaluate the importance of each issue to the campus." Two observations are apparent: the percentages tend to decline with time, and are lower than for self-reported concern.

Figure 7.4 depicts the disparity between students' answers for "how important is environmental issue X to the campus?" versus "how important is environmental issue X to me?" When the bar is positive, it means that respondents felt that the environmental issue is more important to campus than to them personally (i.e., a higher percentage said that the issue was important/very important to UCSC than self-reported it as important/very important). If the bar is negative, it means that more students reported the environmental issue as important/very important to them than to the campus. In 2016, there tended to be more or less equivalent levels of environmental concern that students felt they and campus held, except in the case of agroecology, which was reported to be much more important (by 20 percentage points) to the campus.

By 2019, a marked shift occurs: all bars are negative, indicating that students no longer feel that the campus matches their level of environmental concern, a perception that remains or worsens in 2022 (except in the case of sustainable transportation, which becomes less negative). The largest disparities (greater than 20 percentage points in 2022) between how important issues are to students compared to how they perceive them to be for the campus are seen in renewable energy, water conservation, and waste reduction.

To contextualize these findings, we share a selection of undergraduate comments from the 2022 survey, when we specifically asked students to briefly explain their assessment of the importance of environmental sustainability to them as compared to the campus:

> A lot of things that the campus could be doing are not done because the entire UC system functions to bring in a profit which goes against environmental sustainability.

> Although the school creates an image of caring about environmental justice, I feel like it is only through the lens of less food waste and hardly emphasizes the need to divest from fossil fuels, green infrastructure, and Indigenous practices to maintain the environment.

> Campus or organizations on campus have done a great job of implementing sustainable infrastructure on campus. There are recycling and composting and regular waste bins all over campus facilities. The bathrooms have water conservation flushing in most buildings, and the dining halls offer compostable to-go boxes and utensils that are disposable; they also offer eco-boxes which can be reused time and time again by students. I think having all of these types of things on campus illustrates how much the campus is trying to practice sustainability.

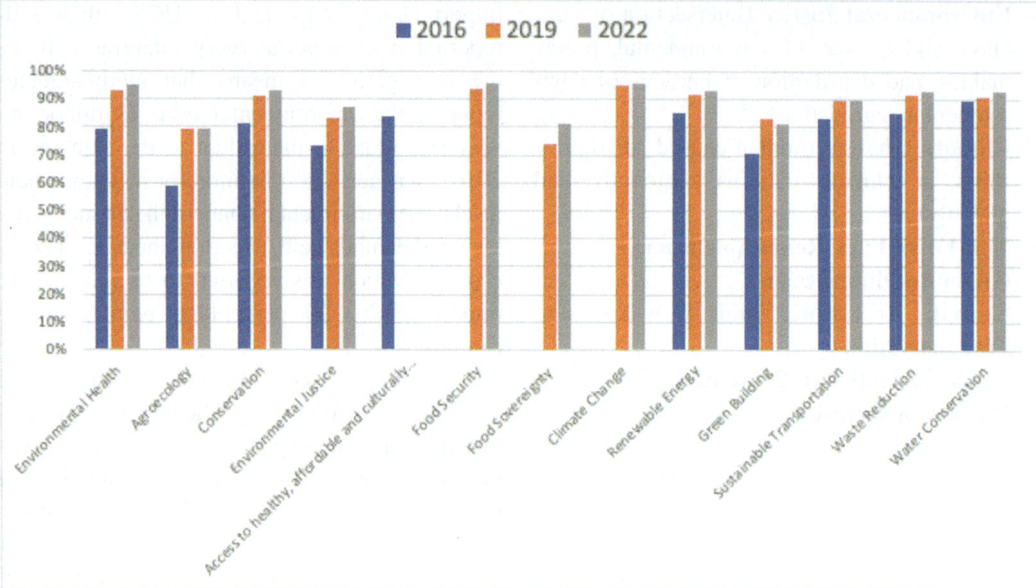

Fig. 7.1 UCSC undergraduates' environmental concerns (% rating each issue as important or very important) in 2016, 2019, and 2022

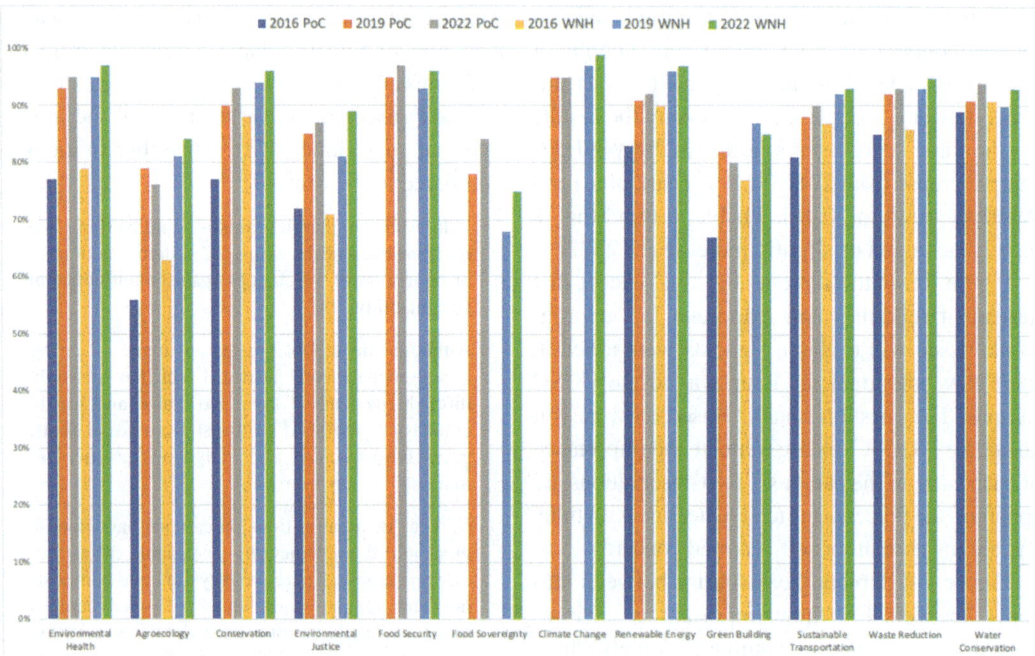

Fig. 7.2 UCSC undergraduates' environmental concerns, by race, in 2016, 2019, and 2022

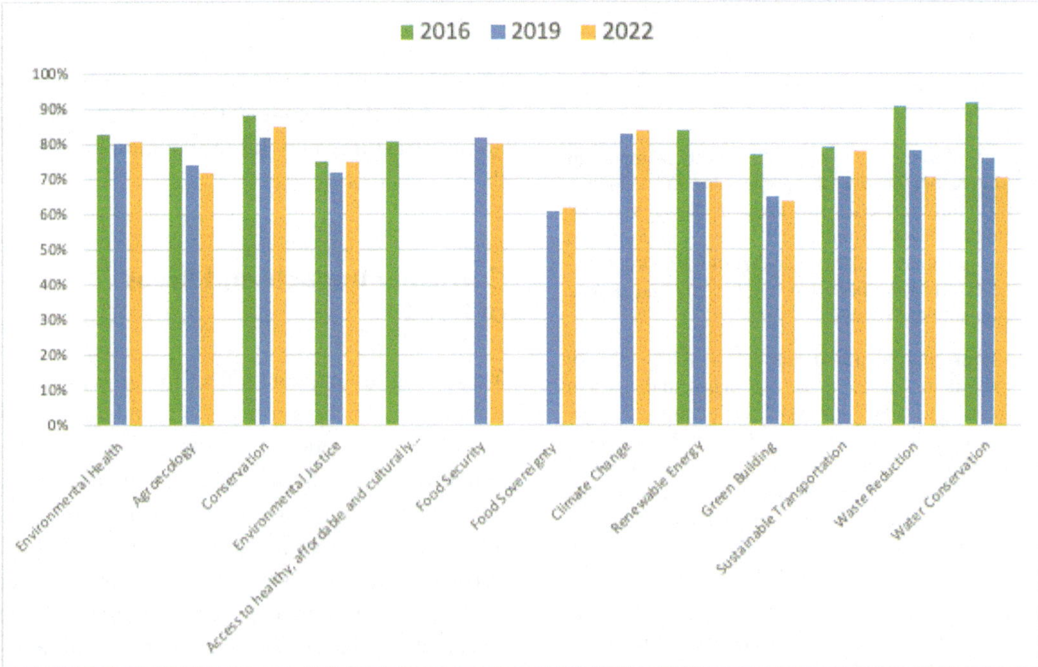

Fig. 7.3 Percentage of UCSC undergraduates who reported that these environmental issues were important or very important to the campus, in 2016, 2019 and 2022

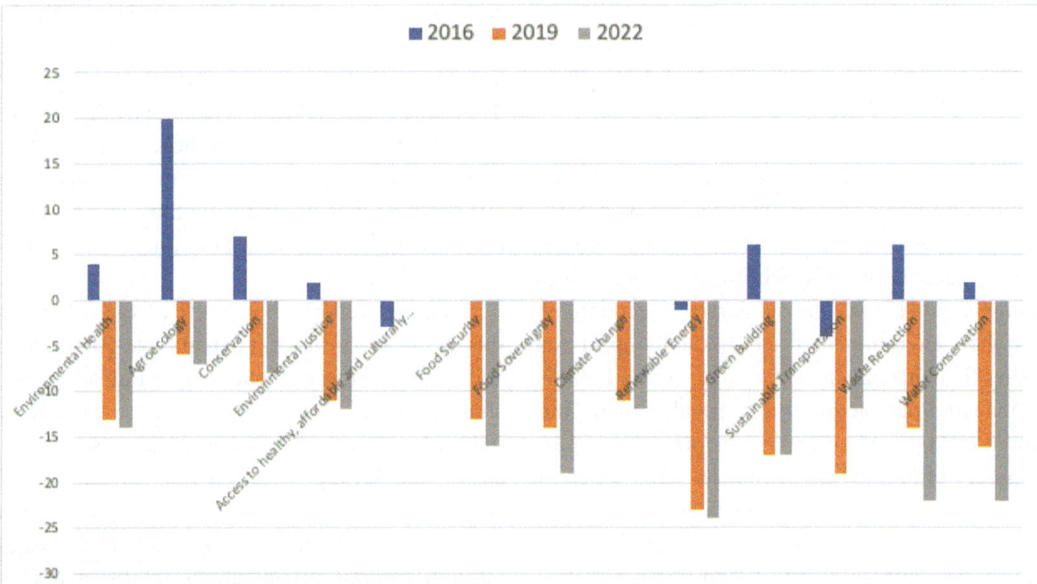

Fig. 7.4 Difference between undergraduates' self-report of environmental concern with their perception of the campus' environmental concern, in 2016, 2019 and 2022

> As UCSC is on fragile land which was stolen from Native Americans and climate change is becoming more and more prevalent, the campus should be prioritizing the environmental aspects of their decisions.

> I feel like environmental sustainability is important to possibly faculty on campus and the students, including myself. However, I do not think environmental sustainability is not as important to the university leaders given that most environmental efforts have been led by students and people who are not in charge.

> I think that environmental sustainability is very important, crucial in fact, to many students and faculty at UCSC, however I don't believe that the structural framework of UCSC as an institution is set up to take it seriously. We need more horizontal leadership; we need to center the voices of the people most impacted by environmental degradation and social injustice; we need to be prioritizing these issues for the present and the future.

> In my experience, it is the faculty that is very passionate and understanding as to how environmental issues intersect through culture, race, and all of our lives. The administration at UCSC and of all UCs fail to even understand the gravity of the environmental problems we face collectively and choose not to listen to the research and knowledge available within their campuses to reach their carbon neutrality goals for 2030, or reduce overall waste and consumption of their campuses. They are reactive not only in their handling of environmental issues and campus impacts, but also in other issues such as race and workers' pay.

As these open-ended responses show, students are much more aware of and engaged with environmental issues, and want to hold their institutions—especially those who market themselves using an environmentalist mantra—accountable for their environmental actions and impacts as well. Students are aware of the contradictions when words and actions are not in alignment, and are attuned to and critical of performative or perfunctory efforts by the institution. This series of quotes also highlights findings in an earlier chapter of this book, which explored different definitions of sustainability and found that a plurality of definitions coexist on the campus. Students who hold a more facilities, resource-centered and infrastructural approach to sustainability can point to efforts to reduce the waste stream, conserve water and energy, etc., while those who adhere to definitions of sustainability that speak to different cultural understandings, trouble historical and systemic inequities, or challenge economic structures may be less complimentary to the university's efforts.

Curricular Impacts and Areas of Need

These longitudinal data reflect the growing environmental concerns of UCSC undergraduates and their perception that this concern is not matched by the campus. One key area in which the university can manifest its commitment to addressing environmental problems is through curricular offerings. We wanted to know whether students were taking environmental- and sustainability-based courses, what specific environmental topics they were engaging with, and if they were gaining an understanding of environmental issues in their general education regardless of major. In addition, we ascertained what undergraduates deemed as key gaps in terms of curricular offerings. We draw on data from the 2022 survey, running statistical tests (chi-square test for dichotomous categorical variables) to determine if differences were statistically significant. We continue to present these data disaggregating students of color and white students given our central focus on understanding and building more racially inclusive environmental educational experiences.

Table 7.1 reports the results of the following survey question: to what extent have you learned about the following issues in your academic courses on the environment or sustainability (through readings, lectures, class discussions, etc.)? The survey specified that students were only being asked about their course work, not including exposure to environmental issues through campus organizations, clubs, and programs. Almost 40% of our students have learned about environmental topics in terms of biodiversity protection, race/class inequities domestically and beyond, and human livability (e.g., food and

Table 7.1 Percentages of UCSC undergraduates who reported learning about a variety of environmental issues "often or very often" through their coursework, in 2022

	PoC	WNH	Int'l	Total	
	n = 1063	n = 568	n = 92	n = 1724	Significance PoC vs. WNH
Race and class-based inequalities in relation to environmental sustainability in the USA	38%	39%	25%	38%	–
Race and class-based inequalities in relation to environmental sustainability globally	36%	34%	20%	35%	–
American Indian/Indigenous understandings of environmental sustainability	27%	29%	17%	27%	–
Non-industrialized countries' approaches to environmental sustainability	20%	17%	18%	19%	–
Biodiversity (ecology, restoration, protected areas, conservation, etc.)	37%	42%	20%	38%	–
Human livability and livelihoods (housing, economics, health) as environmental concerns	38%	37%	23%	37%	–

housing security, community health). We found no statistically significant differences between students of color and white, non-Hispanic students, except in the category of biodiversity, for which the greater percentage for white students was marginally significant at $p = 0.056$. (Given the small sample size for international students, we did not include them in the analysis but provide the results in the table). A clear gap of curricular coverage was apparent with two topics: Indigenous sustainability studies and environmental efforts of non-industrialized nations.

We also asked the following question: have you focused on race and class-based inequalities related to sustainability or environmental in a class research paper, issue/position paper, or creative project? Overall, 42% of the respondents said yes, and 58% said no, with the percentages for PoC and WNH almost identical. We found that 51% of students affiliated with the Social Sciences Division said yes to this question, a statistically significant difference from all the other divisions (Arts 34%; Engineering 36%; Physical and Biological Sciences 37%; and Humanities 41%).

The survey also asked respondents to rate their understanding of environmental issues when they started at this campus and now (Table 7.2). Numbers included are those who answered both the "started" and "now" questions. "Increased" represents those students who started at "good" or below, and increased at least one level. A key takeaway from the analysis is that about 45% of PoC and WNH students increased their understanding of (mainstream) environmental sustainability issues and race and class based inequalities in environmental concerns.

Questions also asked students about their perceptions of curricular needs at UCSC. Table 7.3 summarizes the responses to the question of what students would like to see more attention paid to in their courses. A majority of students want greater coverage of climate change, structural foundations of environmental injustice, and representation of local communities, other countries, and diversity by gender, race and other characteristics. Almost two-thirds of students of color said that there should be more gender, racial, ethnic, etc. representation, which was significantly higher than WNH respondents, and the highest percentage of all results to this part of the survey.

Faculty Perspectives on Environmental Education at UCSC

Given the desire for students to engage in more educational opportunities that link environmental concerns with social justice, and evidence from our own experiences that faculty who research and teach on these issues often do so in relative marginality and isolation in their departments and across campus, we began a project in 2021 entitled, "Identifying Critical Environmental Work and Needs on Campus." The purpose was to create a centralized clearinghouse of faculty across divisions whose research interests broadly relate to understanding environmental issues through themes of power relations, (in)equity, accessibility, epistemology, and social (in)justice, to make their work more collectively visible and create easier access to a shared network. We identified almost 60 faculty and staff doing critical environmentalism work. Serena Campbell, an undergraduate majoring Sociology and Critical Race and Ethnic Studies (CRES), was awarded a Building Belonging Fellowship through UCSC's Institute for Social Transformation to conduct this study.[1] She worked closely with three faculty mentors (the other authors of this chapter) to refine the methodological approach that would allow us to not only better understand current faculty endeavors furthering critical environmental work on campus, but also to support the building of a network to overcome silos and identify obstacles to making such work more effective and collaborative. In Fall of 2021, 58 faculty were invited to take a survey through Google

[1] Before undertaking the research, we verified with the campus' Office of Research Compliance Administration that the work did not necessitate Institutional Review for Human Subjects because the research was being conducted for institutional purposes.

Table 7.2 Changes in understanding of environmental issues while at UCSC, from 2022 survey

		POC	WNH	International	Total
		n = 1052	n = 561	n = 89	n = 1702
Environment and sustainability-related issues	Stayed very good/excellent	17%	23%	14%	19%
	Increased	44%	46%	14%	43%
	No change	39%	31%	71%	38%
Race and class-based inequalities in the history and current debates on environment and sustainability	Stayed very good/excellent	17%	14%	4%	15%
	Increased	44%	46%	21%	44%
	No change	39%	39%	75%	41%

Table 7.3 Topics UCSC students would like to see more curricular attention to, from 2022 survey

	POC	WNH	International	Total	
	n = 722	n = 416	n = 59	n = 1197	Significance POC vs. WNH
More environmental justice topics covered	45%	45%	41%	45%	–
More case studies of different places/countries	54%	56%	59%	55%	–
More perspectives from local communities	57%	56%	52%	57%	–
More gender, racial, ethnic, etc. representation	63%	50%	42%	57%	$p < 0.001$ PoC > WNH
More emphasis on environmental policy and law	54%	52%	44%	53%	–
More focus on forms of resistance and mobilization	44%	50%	37%	46%	$p < 0.05$ WNH > PoC
Deeper analysis of the structural and historical roots of environmental injustice	54%	60%	47%	56%	$p < 0.05$ WNH > PoC
More emphasis on climate change or the climate crisis	58%	59%	54%	58%	–

forms with a combination of closed- and open-ended questions asking about their: background; training and experience; research and scholarly work; courses and teaching; other initiatives, projects, and co-curricular activities/programming; and need for institutional support. We received 39 responses, or a 67% response rate. Serena synthesized the results of this survey, and then followed up with 30 faculty members for an in-person interview during their office hours. In this follow up interview, she focused on eliciting what kinds of targeted policy and actions could be taken towards critical environmentalism on campus. In total, we were able to gather information from 56 UCSC faculty from across departments and divisions.

For the most part, through the survey data and follow-up interviews, faculty were appreciative of the opportunity to discuss their projects and the changes they would like to see on campus. Some were skeptical that any institutional changes could be made that would promote critical environmentalisms, but others were excited about the idea and gave suggestions for what they would like to see get done. Many were frustrated about how little the campus appeared to be invested in this work, which they took as a sign that the university does not see critical environmental work as a priority. However, they had not lost faith that changes could be made. The conversations were extremely informative and the respondents brought interesting ideas or suggestions for what support could be beneficial. Just as undergraduates voice an interest in more equity-centered educational spaces to engage environmental concerns, faculty also want to catalyze and make visible more critical environmental sustainability praxis at UCSC, and have diverse ideas about what form it can take.

Faculty members articulated multiple ideas and strategies to promote more critical environmental work at UCSC:

- Change the incentive structure to support faculty to undertake critical environmental work both on- and off-campus. As one faculty member articulated, "There's not really a lot of incentive for faculty to be involved in either academic senate work that is related to fighting the climate crisis or even grassroots efforts on campus, they're not rewarded by course releases or anything like that." Seven respondents (12.5%) mentioned the lack of incentives to engage in critical environmental work, and 31 faculty members (55%) specifically named course releases as a needed incentive to give time to faculty to pursue different work/research projects.
- Create a webpage, app and/or database of various critical environmental efforts happening on campus so that people can easily stay up to date. A majority (62.5%) of faculty respondents mentioned this point, such as one professor who said, "A web presence, some central list of people who are working and what they are working on specifically but whose work kinda fits under this broad umbrella, I think would be the first and most obvious thing that would be super helpful. Some place that's easy to go to that would be identifiable off campus… being able to point out cross departmental or divisional initiatives."
- Foster interdisciplinary collaboration on research, teaching, and socializing/networking across disciplines. Giving people the opportunity to meet others across divisions with similar interests has the potential to lead to collaborative projects and perhaps pursue funding geared towards interdisciplinary and transdisciplinary efforts. "The fact that these academic territories are carved up means that often there is not a lot of communication that is centralized… and that means in certain ways there's a limitation on the scope of collaborative possibilities and communication," said one professor. One mechanism to achieve this would be to create a critical environmental cluster, center or institute on campus, an idea which was mentioned by 45% of respondents. As one professor noted, "I would love to see more avenues for connections across programs and departments within the university. I think we often get siloed into our own programs or departments, but I just think that if we were able to work more collaboratively

we would provide richer experiences for students. I do think that to be able to do that, the programs within themselves need to be well staffed because otherwise we would just stay in the state of being overextended and barely getting by and not having the time to create."

- Create a community around critical environmentalism through routine get-togethers to present works in progress and meet people who have the same interests—faculty wanted spaces to bring together people in different roles on campus, from undergraduates to professors. This is particularly needed since "COVID has really limited the visibility of all of us with each other, so the first thing I think of are social things." More environmental events (64%) and critical environmental networking opportunities (46%) were commonly articulated needs.
- Revisiting curricular offerings was another suggestion, specifically the opportunity to co-teach classes to foster more interdisciplinary engagement (37.5%) and also to hire more diverse faculty with this expertise to offer more courses (7%).
- Expand and make visible resources and opportunities for undergraduates and graduate students to get involved in critical environmental work. "The students of color that are engaged in conversations around critical environmentalisms and sustainability are very smart, very engaged, and they have some really good ideas about what could be done to engage their experiences more," said one faculty member. Another stated, "Student success is a really essential component of what we should be and I feel like we are not doing enough to ensure student success." Over 55% of respondents wanted additional and more visible opportunities for students, and another 12% emphasized the need to focus on students of color.

Unsurprisingly, three-fourths of all respondents plainly stated that more funding is needed for critical environmental efforts to undertake the projects articulated above, from hiring the staff person to maintain a central clearinghouse or information hub/database, to provisioning course releases to enable faculty to undertake this kind of research, community engagement and advocacy. As one professor summarized, "If there can be ways of overcoming institutional barriers to free people up in meaningful ways, not just invite people to get together or discussions on top of everything else because everyone is just so busy with teaching, with research, with administrative obligations at the departmental, divisional, and senate level." Some faculty shared comments critiquing the institutional structures that create a lack of incentives, support and collaboration around critical environmentalisms. The following faculty quotes capture this sentiment:

> We need more resources and support for these institutes and offices. Less money for the regents and president's office and more money for universities and institutes and offices.

> I think that the university needs to listen to its constituents, it needs to listen to its students and faculty, and decisions around planning need to not be made by a couple of executive vice chancellors in closed rooms with contractors. The institutional support really needs to be around transforming the whole university and centering student staff and the broader community, the Indigenous community on which UCSC stands. I think the institutional support that I envision requires a full transformation of the university.

> Conceptualizing and making sense of critical environmentalism on a campus such as UCSC requires intentional restructuring and institutional framing in order to begin to upset the incredibly rooted white supremacist history of environmentalism on this campus.

Conclusion

The "Beyond the Bullsh*t" conference organized four panel presentations featuring current undergraduate and graduate students and alumni. The "Young Activists Panel" brought together Ayo Banjo, former UCSC Student Union Assembly President and member of the CA/HI Executive Board of the NAACP; Zennon Ulyate-Crow, founder of the Student Housing Coalition and

Transportation and Public Works Commissioner for the City of Santa Cruz; and Lisette Jones, Field Representative for Assemblymember Mark Stone. They reiterated that climate change is the defining issue and fight for Generation Z and Generation Alpha, and addressing the challenges ahead will take greater intergenerational capacity and coalition building. Each drew upon their experiences at UCSC and the surrounding community. Zennon highlighted the neo-Malthusian discourses that still pervade environmental spaces in the City of Santa Cruz, a place that is "only getting whiter and wealthier" and one that resists building "denser and more diverse" to accommodate everyone. Ayo recalled often being the only Black person in his classes, and how he worked with Asian and Latinx students to undertake a lithium-ion battery project, which led to a meeting with climate tech investors. Ayo did consulting work to connect Black engineers with climate jobs in Santa Cruz, and organized an equity summit to address issues of representation and diversity in the industry. Lisette, an ENVS major who graduated in 2021, spoke about the predominantly white spaces she encountered at the university and among policy writers in state government. Canonical texts used in her ENVS 25 course, she noted, were authored by racists and white supremacists, a fundamental failure that needs to be addressed. For these panelists, on the forefront of addressing contemporary challenges around equity, liveability, and governance, the "status quo is broken", and "we need to fight those power dynamics out there [that oppress historically marginalized groups], starting with UCSC."

Longitudinal survey data presented in this chapter reflect the high and increasing levels of undergraduate concern since 2016 across a range of environmental issues, a concern that they do not feel is matched by the positions and efforts of the campus. We found that only around 40% of the student body in 2022 are often engaging with environmental content in their courses, and about the same percentage report increases in their environmental understanding during their time at the university. Glaring gaps in the curricula consist of Indigenous environmental knowledge and culture and a global scope of environmental efforts, especially in non-industrialized countries. While reports of existing course content and learning gains around environmental concepts did not differ between students of color and white students, a key finding of the research is that almost two-thirds of students of color said that there should be more gender, racial, ethnic, etc. representation, which was significantly higher than WNH respondents. This is a clarion call for urgent changes that students want to see in how environmental issues are framed, taught, and addressed.

The work of the Identifying Critical Environmental Work and Needs on Campus project shows that the students are not alone in these calls—that their demands reflect those of faculty who also want to see more support for equity- and justice-centered approaches to addressing climate change and other threats. The project involved identifying and then surveying and interviewing faculty at UCSC who engage in critical environmental work that they do not feel supported by the institution to do. Respondents indicated that they lacked time and opportunities that would allow for them to engage in critical environmental work to the extent that they would like. This includes opportunities for interdisciplinary networking, collaboration, and co-teaching; student engagement; new critical environmental hires; and incentives to engage in critical environmental work. They believe that the university's focus on supporting and funding "hard" sciences ignores the importance of social justice and equity-centered approaches to addressing contemporary environmental challenges such as the uneven impacts of climate change. This imbalance also leaves faculty who do critical environmental work undervalued and overextended. Respondents stated that the campus can bridge this gap by: establishing a formal critical environmental center, institute, or cluster on campus; building a critical environmentalism network that hosts events; holding social events with critical environmental themes; offering more funding for critical environmentalism work;

setting up a webpage for faculty and staff doing critical environmentalism work; incentivizing efforts through course releases and more administrative support; and funding graduate student researchers and staff positions. Overall, respondents showed that there is a great deal of passion for critical environmental work on campus, but resources and support need to be built so that students, faculty, and staff can engage in it and help shape UCSC to become the global leader in critical environmentalism that they believe it can be.

References

Bleemer, Z. (2022). Affirmative action, mismatch, and economic mobility after California's proposition 209. *The Quarterly Journal of Economics, 137*(1), 115–160.

Dare, K. (2021). *Propelling UCSC undergraduates of color in environmental studies through mentorship* (Senior thesis). https://transform.ucsc.edu/building-belonging-fellows-2020-2021/

Hernandez-Jason, S. (2020). *UC Santa Cruz ranks again as a top university for social mobility*. UC Santa Cruz Newscenter. Retrieved December 20, 2022, from https://news.ucsc.edu/2020/09/usnwr-rankings.html

Leonhardt, D. (2015). California's upward mobility machine. *New York Times*. https://www.nytimes.com/2015/09/17/upshot/californias-university-system-an-upward-mobility-machine.html

Reti, I. H, Vanderscoff, C., & Rabkin, S. (2020a). *Seeds of something different: An Oral History of the University of California* (Vol II). Retrieved December 20, 2022, from https://escholarship.org/uc/item/4wn6f291

Townsend, P. (2015). *The birth of UC Santa Cruz: Audacious and academic*. UC Santa Cruz Newscenter. Retrieved December 30, 2022, from https://news.ucsc.edu/2014/12/birth-of-ucsc.html

University of California Information Center. (2022a). *Disaggregated data*. University of California Regents. Retrieved December 30, 2022, from https://www.universityofcalifornia.edu/about-us/information-center/disaggregated-data

University of California Information Center. (2022b). *Undergraduate admissions summary*. University of California Regents. Retrieved December 30, 2022, from https://www.universityofcalifornia.edu/about-us/information-center/disaggregated-data

University of California Regents. (2015). *Accountability report*. Retrieved December 30, 2022, from https://accountability.universityofcalifornia.edu/2015/

University of California Regents. (2022). *Accountability report*. Retrieved December 30, 2022, from https://accountability.universityofcalifornia.edu/2022/

Reti, I. H, Vanderscoff, C., & Rabkin, S. (2020b). *Seeds of something different: An Oral History of the University of California* (Vol. 1). UC Santa Cruz Library. UC Santa Cruz: Regional History Project Oral Histories. Retrieved December 20, 2022, from https://escholarship.org/uc/item/68v4q9sf

Flora Lu is a Professor of Environmental Studies at the University of California, Santa Cruz and an ecological anthropologist specializing in conservation politics, environmental justice, and political ecology. She served as the Provost of College Nine and John R. Lewis College from 2014 - 2023, during which time she helped establish the People of Color Sustainability Collective (PoCSC) and the Apprenticeship in Community Engaged Research (H)ACER. She was a Postdoctoral Fellow at Stanford University and an Assistant Professor in Ecology at the University of North Carolina at Chapel Hill before joining the UCSC faculty in 2008.

Emily Murai is a Lecturer at College Nine and the Environmental Studies department at the University of California, Santa Cruz. She teaches courses on critical reading, writing and thinking, writing-across-the curriculum/writing-in-the-disciplines (WAC/WID), and in the environmental humanities. She holds a Ph.D. in Human Geography from the University of Minnesota, an M.A. in Sociology from the New School for Social Research and a B.A. in American Studies and Psychology from UCSC.

Serena Campbell is an undergraduate student at the University of California, Santa Cruz majoring in Sociology and Critical Race & Ethnic Studies who will graduate in 2023. She participated in the project, "Identifying Critical Environmental Work and Needs on Campus," and wrote an award-winning senior thesis based upon the research she helped conduct.

Hillary Angelo is an Associate Professor of Sociology at the University of California, Santa Cruz (UCSC). Her work offers a social-theoretical perspective on socio-ecological questions through historical and contemporary research on urban greening, sustainability planning and policy, infrastructure, and climate change. Before completing her PhD, she worked for five years with the New York City Department of Parks & Recreation on issues of participatory design, immigration, and public space use.

Part III

Community-Engaged Critical Sustainabilities

8

Developing a Praxis of Loving Relations: Lessons from a Community-University Partnership that Centers Undergraduate Research and Learning

Linnea K. Beckett
and Michelle Hernandez Romero

Abstract

This chapter outlines the Apprenticeship for Community Engaged Research or (H)ACER program at the University of California, Santa Cruz. It explores how co-author Michelle Hernandez, while as an undergraduate, drew from decolonial feminist scholar Maria Lugones' theorizing on loving perception and world traveling to develop her own praxis of community-engagement as a research apprentice in an immigrant-led, one-acre community garden. Michelle describes the practices that she used to build loving relations with the gardeners and how this in turn supported her own academic journey. This chapter is designed to open up educators' imaginations on how higher education programming can create dynamic and diverse pedagogical spaces for students and community members to learn from each other.

Keywords

Critical community-engaged research · Critical community learning · Anti-racist higher education programming · Popular education · Decolonial feminisms · Community gardens · University-community partnerships · Community building · World traveling

Introduction

> My first time meeting Don Luis,[1] I immediately felt this warmth in my heart. The same warmth you feel when you reminisce of home. He and his family remind me of my family so much. The way they speak, the way they refer to things, the things they laugh about, the things they eat. It feels like having dinner at my abuelita's house on a Sunday night.

Michelle jotted down these field notes in 2018 one day after visiting the Calabasas Community Garden. She writes as an undergraduate student who has been working in the Apprenticeship for Community Engaged Research or (H)ACER program for about a year as a research apprentice in the community garden. She has been studying why gardeners choose to participate in the space and over the past few months, gathering interviews with gardeners and field notes in the garden.

L. K. Beckett (✉) · M. Hernandez Romero
University of California, Santa Cruz, CA, USA
e-mail: lbeckett@ucsc.edu

[1] We have used pseudonyms throughout to protect the identity of the gardeners.

F. Lu, E. Murai (eds.), *Critical Campus Sustainabilities*, Sustainable Development Goals Series,
https://doi.org/10.1007/978-3-031-30929-8_8

As she researches, she also helps out in the garden with overall maintenance and upkeep of the space, following the gardener's directions and lead. This field note describes her first time meeting one of the gardeners and conjures up loving visions of family, home and the *cariño* that one feels when sharing food together, as a family.

As Michelle deepened her relationship with the garden and with the families working in the garden, her field notes started to reflect her own experiences and relationship with the space. She began to share how the space called forth her own memories of her family and home. She continues, sharing a story about an interaction she had with Leonardo, Fernanda and Don Luis. Don Luis is in his 80s and is the father to his adult children, Fernanda and Leonardo. Together the family tends to two plots in the garden.

> It was a Wednesday morning. I went to the community garden to do some work and drop off some stuff. Leonardo, Fernanda, and Don Luis showed up to work on their plots. I used to always see them on Wednesday, and I always looked forward to it. I walked over to their plots. Don Luis immediately offered me seeds, "Mira Michelle, ten. Para los elotitos." He gave me a handful of red corn. "Que caray Michelle, Fernanda nos dijo que eres nuestra vecina!" Leonardo mentioned, acknowledging that we are from the same town in Michoacan, Mexico. We spoke about the food, they knew what Morisqueta is. We spoke about el Parque Nacional and the crazy amazing kids that dive in the dangerous rocky rivers for money. We spoke about las monarcas, and Dia De Los Muertos when they fly through the city on their way to our mountains. We spoke about el Mercado de Antojitos and how it smells like atole de grano, uchepos, y corundas. We spoke about where we lived, we found out our families live only a few blocks away from each other. We were remembering and laughing as Don Luis, Fernanda, and Leonardo tended their plots. Don Luis is very gentle, he moves slowly and you can almost hear him talking to the plants under his breath. It reminded me of how mi abuelito used to do it. He was always quiet in his huerta, but if you were close enough you could hear him singing.

As Don Luis handed Michelle some corn seeds that he had brought from Mexico to grow in the garden, he also mentioned that his daughter Fernanda had told him that she had learned Michelle's family lived just around the corner from his home in Michoacán. Neighbors in the same town in Michoacán, they now found each other thousands of miles away from their home town in a community garden in Watsonville, California. They reminisce together about the food, the rivers, ceremonies, the smells of food and the vibrant cultural practices of that region. Remembering and laughing, Michelle sees her grandfather in Don Luis's movements as he sings to the plants and tends to his plot in the garden.

This is a story of finding love and healing in unexpected places. Michelle didn't know, as she signed up to be a (H)ACER research apprentice in her undergraduate studies, that she would, in some ways, find her way home to her grandfather's farm. This unexpected encounter fueled a deep and long love Michelle developed for the Calabasas Community Garden and for the reflexive research practices that supported her to recover and remember her roots. As she worked with families and the garden, she would continue to remember who she was and where she came from. These powerful relationships and cultural practices would sustain her for the rest of her time as an undergraduate student.

Established at John R. Lewis College at the University of California, Santa Cruz (UCSC), the (H)ACER program bridges community-engaged scholarship (Gordon da Cruz, 2017) with undergraduate education. It offers undergraduate community-based programming and a variety of activities ranging from a one-day workshop on food justice, or workday at the Calabasas Community Garden, to a year-long research apprenticeship in the garden. Through these experiences, undergraduates learn with each other and with communities about pressing social, environmental, educational and cultural issues that both students and communities may be facing, and explore creative ways to respond to these issues. Grounded in tenets of popular education and feminist pedagogies, the (H)ACER Program centers relationships—student and community—and seeks to create space and time for people from diverse walks of life to share and learn from each other.

The Calabasas Community Garden is an acre-sized community garden situated on a plot adjacent to Calabasas Elementary School, which is a semi-rural public elementary school in Watsonville, CA. Surrounded by apple orchards, strawberry fields and a rural airport, Calabasas Elementary School serves a 97% Latinx student body, all of whom qualify for a free and reduced lunch. On an average year, the garden supports approximately 35 families, providing 15 × 15 ft plots for families to cultivate a large *milpa*, a squash communal field. The community garden privileges relationships by design, with contractual agreements requiring attendance at community meetings and community hours instead of monetary transactions. The elementary school after school program runs in the garden and teachers also have plots interspersed among gardener's plots.[2]

Michelle was one of the first students in the (H)ACER program and as she moved through the program, sharing her reflections, she provided important insights for future envisioning, theorizing and programming. After her two years as a student in the (H)ACER program, Michelle became its Community Outreach Coordinator and coordinated undergraduate visits to the Calabasas Community Garden. In this paper, I (Linnea) share the theory and structure of the (H) ACER Program, designed as an anti-racist, anti-colonial intervention to support first-generation, BIPOC students at UC Santa Cruz. I then draw from Michelle's past and present voice to share some of the ways she theorized her experiences in the (H)ACER program and garden. In offering her experiences alongside a descripton of (H) ACER, Michelle and I hope to breathe life into the academic explanations of (H)ACER. Through her work in (H)ACER, Michelle developed a praxis of loving relations. She drew from decolonial feminist scholar Maria Lugones' (2003) theorizing on loving perception and world traveling to hone and articulate her own praxis of community-engagement. I draw from her descriptions of community engagement to articulate her praxis of entering into, following and caring for her relationships with the community gardeners as an undergraduate student researcher.

[2]We will expand on the history of the Calabasas Community Garden in the following chapter.

This chapter is designed to open up our thinking about the possibilities and limits of university programming to support justice-oriented work. The stories are non-exhaustive and were not collected through a research design. They are not intended to be generalizable to all of the students within the program. Instead, these stories should propel our imaginations as educators, advocates, and justice dreamers to explore how to create spaces in higher education toward justice in ways that also implicates the university itself in these justice struggles. The history of modern universities in the US show us that universities have and continue to benefit from Indigenous dispossession, slavery, racial capitalism and white supremacy (la paperson, 2017; Stein, 2018, 2020; Tuhiwai Smith, 2021) and the knowledge systems upheld and revered within universities promote and maintain coloniality more broadly (Quijano, 2000; Lugones, 2007). Given these complicated and violent histories, how can we create pedagogical possibilities and relationships both on and off campus toward life-affirming practices and relations? In other words, how can we envision higher education programming that seeks to not only generate more inclusive spaces, but that moves in the direction of transforming the university to better serve students from non-dominant communities?

The (H)ACER Program: An Apprenticeship in Community-Engaged Research

The Apprenticeship in Community Engaged Research or (H)ACER began in 2018 as a pilot program to support UCSC's growing Latiné/Latinx student body, first-generation students, transfer students, and students of color more broadly. (H)ACER bridges academic and co-curricular programming, which allows it to offer critical service-learning opportunities and research apprenticeships side by side (see Beckett et al., 2022). Through deep partnerships with a select few community groups in Santa Cruz County, (H)ACER creates multi-faceted, place-based programming. The Calabasas Community

Garden is one of the sites of this programming. Undergraduates can volunteer to participate in workdays at the garden. They can work over the course of a quarter with third, fourth and fifth grade students from the elementary school in an after-school program in the Calabasas Community Garden, facilitating games and activities with the elementary school students. Or, students can apply to become year-long researchers in the garden, in which they attend monthly community meetings, write extensive field notes and conduct interviews with community gardeners as part of an ongoing participatory research project which is designed to visibilize the learning, specifically STEM learning, occurring in the garden.

Additionally, (H)ACER runs programming at UCSC that showcases the gardeners and the community garden. The garden manager and gardeners sit on panels sponsored by John R. Lewis College at least once a year, often multiple times for different classes. They discuss food justice, food sovereignty and garden-based education. These events showcase the funds of knowledge (González et al., 2006) of gardeners and highlight the "community cultural wealth" (Yosso, 2016) as important aspects of student learning. It also provides an opportunity for the gardeners to develop their leadership skills and further strengthen the garden's community and governance.

It is through these multi-layered activities and multi-sited relationships that (H)ACER supports undergraduates to be in the garden at varying levels of participation. Students interested in the research apprenticeship, for example, most likely learned about the garden through PRAXIS, Alternative Spring Break, and the Calabasas After School Internship, which are programs run through (H)ACER that integrate undergraduate participation in garden activities. Through these programs, students deepened their connection to the garden and sought other ways to stay connected. At John R. Lewis College, where co-curricular, out-of-classroom learning and academic learning are equally valued for students to receive a liberal arts education, the diverse learning opportunities of (H)ACER seamlessly integrate. In addition to the experiential learning and research opportunities, undergraduates can take academic courses on community-engaged research methodologies, which provides an important theoretical base for critically reflecting upon power dynamics, being in community, and learning about the politics of knowledge production. Through an apprenticeship model and a set of principles (outlined below) that guides the curriculum, students are trained to develop their ethnographic sensibilities and think meta-cognitively like researchers, which supports critical consciousness raising (Freire, 2021).

Developing Ethnographic Sensibilities

Beckett et al. (2022) describe the founding principles programmatic structure of the (H)ACER Program and I summarize these principles here. We draw from the concept of apprenticeship (Lave & Wenger, 1991), which privileges learning with others (peer, near-peer and intergenerationally), learning by observing, and learning in collaboration. We see learning as situated in environments with artifacts (Engeström, 2001) and co-constructed (Vygotsky, 1978) among peers, elders, youth (Lave & Wenger, 1991). In this arrangement, professors are not experts disseminating knowledge, but facilitators, along with community members, undergraduates and other participants in knowledge co-construction.

Through opportunities that span from one day volunteering, to one-, two- or five-unit courses, to paid research internships, undergraduates learn about and practice ethnographic methods. Through these classes and experiential learning opportunities, (H)ACER students learn *ethnographic sensibilities*, which includes learning how to observe, listen, reflect, describe, and analyze social, cultural, historical phenomena and their own positionality. As students develop these sensibilities, they also practice reflecting upon their own identities and experiences, highlighting their own ways of knowing, which they then think with alongside academic knowledges. Students reflect on how they enter into and build relationships with the environment and communities in which they are situated. As they describe these experiences and their intentionality in building these relationships, we ask students to theorize their own ethics of community engagement. These discussions and assignments are embedded in ongoing conversations about (H) ACER's long-term responsibilities to the Calabasas Community and the garden.

(H)ACER's Guiding Principles

The program's guiding pedagogical philosophy is based in problem-posing education (Freire, 2021), the assumption that undergraduate students and community members are knowledge holders and producers, and through a collective and reflexive study, individuals can develop deeper understandings of themselves and their historicity.

(H)ACER's philosophy recognizes that all of us—community members, educators, and students—are conditioned by intermeshed (Lugones, 2003) oppression, but we are "not determined" (Freire, 2000). In other words, these structures are neither complete nor deterministic. Through critical, collective study, we can learn about the world together, imagine and practice the creative responses to the hegemon. Aligned with critical community-engaged research, our approach relies on community-based funds of knowledge (González et al., 2006), recognizing that people who are adversely affected by oppression are the most well positioned to envision responses to said oppressions (Freire, 2000).

Part of our program examines the racial colonial formation of US higher education (e.g. Stein, 2020; Wilder, 2013) and we continually ask how we can participate in anti-racist and anti-colonial learning, given these constraints (la paperson, 2017; Harney & Moten, 2013). As a program, we link university resources to community-initiated efforts toward dignity, health and well-being, as part of our ethical commitments as scholar-educators at a public university towards advancing social justice. We recognize collaborations and solidarities are ever-evolving, often contingent (Gaztambide-Fernández, 2012; Tuck et al., 2014) and require ongoing care to not reproduce university-community racial colonial power dynamics. (H)ACER only has a few long-term partnerships and these partnerships guide the student programming.

Table 8.1 outlines the program themes and aims of (H)ACER's curricular, theoretical, and pedagogical approaches to training undergraduates as community-engaged learners.

Toward an Ethic of Loving Relations

The philosophical and pedagogical structure of (H)ACER outlined thus far, provides important background for how Michelle and other students explored and theorized their community-engaged work as undergraduate student researchers. In the inaugural cohort of (H)ACER, Michelle and another student spent a lot of time discussing and theorizing a kind of radical relational ethic that they saw emerging in their practice in the garden. They began by writing about it in their field notes. Through shared readings and discussion of those field notes in a research group, the students landed on the concept of an *ethic of loving relations* as a guiding principle that helped describe their experiences in the garden.

During a research group meeting, our team unpacked Maria Lugones (2003) concepts of *loving perception* and *world traveling*. In her book, *Pilgrimages/Peregrinajes: Theorizing Coalition Against Multiple Oppressions*, Lugones describes ways in which people can come together across difference to create coalition. She describes a practice of loving perception, which may allow for a kind of radical relationality to support people coming together and building solidarities. She explains, social fragmentation "keep[s] our lines of resistance away from each other" (Lugones, 2003: 85) and holds us back from traveling into each other's "worlds". She suggests that in order to overcome this social fragmentation and break from the logic of oppression, we need to be able to relate to each other and identify across our worlds of difference. We need loving perception to be able to "travel" into another's

Table 8.1 Principles and commitments that informed the design of the (H)ACER Program

Program Themes	Detailed Program Aims
(1) Racial colonial formation of US higher education and our role as educators and students	(1a) Examine racial colonial histories of US higher education, with a focus on Santa Cruz County and UC Santa Cruz.
(2) The racial colonial formation of US higher education and implications for knowledge production.	(2a) Through a historical and theoretical study of the formation of the university, explore how racial colonial university is part of a larger project in producing colonial/modern onto-epistemologies. (2b) Study critical, feminist and decolonial interventions into traditional social science research. (2c) Describe how we might use research to support community organizations and emergent community-based collaborations.
(3) Pluralize and trouble ideas of "justice" in social justice research	(3a) Recognize how different groups, organizations, and institutions utilize the term 'justice' in very different ways. (3b) Understand how 'justice' and 'social justice' can be diversely operationalized for domesticating and liberating ends.
(4) Critical qualitative research methods	(4a) Study critical, feminist and decolonial social science research methodologies. (4b) Develop skills to conduct qualitative research, i.e. interviews, participant observations, and facilitate focus groups.
(5) Ethnographic sensibilities	(5a) Develop practices of observing and listening. (5b) Study relationship building in diverse cultural contexts. (5c) Utilize inductive, interactive and recursive data collection and analytic strategies to build theory. (5d) Develop consciousness and attentiveness to cultural context.
(6) Positionality and ethical commitments	(6a) Articulate ethical research practices. (6b) Develop capacities to 'be with' contradictions as a form of engagement and provocation toward dialogue (Lugones, 2003). (6c) Explain how theory can be experience-based and "home made," meaning that one's position, history, and relationships matter (Lugones, 2003; Moraga & Anzaldúa, 2022).

world and connect. Loving perception stems from an understanding that

> [w]e are fully dependent upon each other for the possibility of being understood and without this understanding we are not intelligible, we do not make sense, we are not solid, visible, integrated; we are lacking. So traveling to each other's "worlds" would enable us to *be* through living each other (Lugones, 2003: 86).

Through loving perception and world traveling, we can begin to see ourselves as interdependent in resistance, building relationships and solidarities in the process. World traveling necessitates a desire to relate, an openness to break through "ready-made sense" and understand meaning making to be embodied, and situated in relation. World traveling with openness, love, and faith acknowledges the capacity of the relation as an epistemological space. This relational, collective and active "making the world in which we live intelligible to ourselves and to each other – in other words, teaching ourselves" (Alexander, 2005: 6), is the root to imagining that another world is possible.

Practicing an Ethic of Loving Relations

Michelle resonated deeply with this theoretical frame and in October of 2019 she used this frame to describe her own practices of entering into the garden and building relationships with the gardeners. Through this section, I present Michelle's description of how she put into practice her theorization of an ethic of loving relations. I first describe how she entered into the garden, followed the community gardeners' lead, and found her way to being, as she states, "truly accountable" to the gardeners as a researcher in the space. Throughout, she provides clear practices of loving relations, describing in concrete terms how she practiced this ethic as a research apprentice.

Entering in Relation

Upon arriving at the garden, Michelle knew the community she encountered was not her own home community, but at the same time, she shared many similar life stories and cultural practices as an undocumented Mexican immigrant. Through her prior experience in a psychology research lab and readings with (H)ACER that described the extractive history of researchers in communities of color and power dynamics of community and university partnerships, Michelle knew that she did not want to participate in research that was, as she described, "extractive and one sided."

When asked to reflect upon her approach to the gardeners, she used the Spanish verb *acomedar* (*se acomedida*, to be accommodating) to describe how she positioned herself in the garden and how she began to make connections. She described it as "showing up in a loving way is like showing up with that desire to want to be of service and want to contribute to the positive energy that is already flowing and being respectful of everyone's stories and knowledges and just being receptive to what is being given to you."

When she would arrive at the garden, she would not forefront her research agenda or walk into the garden and remain at the periphery, waiting for an interaction or to start writing up field notes. Instead, she would walk into the garden and present herself to the gardeners. She would say hello and then ask, "How can I help?" Searching for a way to be useful, she would connect with the gardeners, learning about their current tasks and supporting their work activities. Positioning herself in a supportive role allowed Michelle to be with the gardeners and to learn what they were doing in the garden, which opened possibilities for her to also learn more about the gardeners, their perspectives and experiences.

Reflecting upon her upbringing, Michelle described *acomedida* as being there for everyone else. "Coming in and thinking of everybody else's well-being." For example, walking into the garden and looking for the elders:

> Going and saying hi to everybody. Asking the elders what they are doing and then going to see the different things that are happening and where you can help. It is like showing up in the same way that I show up to a family party or to my house every time I visit my mom. I go in, I say hi to everybody, see how everybody is doing, ask about what I can help with and then have conversations. There is always something to talk about, random things can come up. It is really just being present and there for everybody… making sure everybody is comfortable.

Michelle explained the concept of *acomedida* through her experiences with her family. She viewed the gardeners in the same light, as kin. Relating to the gardeners as family facilitated Michelle building a connection with the gardeners on a deep level rooted in respect and love.

Following the Community Gardener's Lead

Through her studies with (H)ACER, Michelle was critical of researchers and others "coming into different communities and assuming that we know what they need without talking to them." She explained that this creates hierarchies and "savior complexes", which she knew she did not want to be a part of. Instead, she focused on "being consistent and having mutual relationships." This could be put into practice in a myriad of simple practical ways. One example Michelle provided was describing how she developed a communication style with Yolanda, the Calabasas Community Garden Manager that extended beyond her own work-related schedules.

She learned that Yolanda's schedule was different from her university schedule and she needed to be conscious of and accommodate Yolanda's schedule, even if it was different from hers. She explains, "I realized Yolanda's schedule was really different than the schedules you would get here in the university so…if she has to call me at 10:30 at night, that is fine because that is the only way." She adjusted her schedule to communicate with Yolanda and followed her lead, accommodating her time to support the collective endeavors of the community garden. Michelle attuned to the needs and schedules of Yolanda and the gardeners more broadly, finding space and time to be in relationship with them on their terms. This attunement allowed her to relate to and enter into the gardener's "worlds" in mundane, yet important ways. Through these interactions, Michelle built trust with the gardeners, expressing to them through her actions that she was there to follow their lead and support their work on their terms.

Being Honest and Accountable to the Gardeners

As a researcher and agent of the university, Michelle understood she wore two hats. One as a student working alongside the gardeners in the garden, but also as a researcher, collecting data that focused on how people were describing the function and value of the garden in their lives. As she entered the garden, she foregrounded the former, however her role as a researcher and collecting data could also reproduce the same extractive practices she sought to avoid. She had a strong ethical compass and was clear that "[t]hese types of collaborations are really only possible by building strong, consistent, and honest relationships with the community, and being conscious of where you stand at all times and always aware of who you're truly accountable to." Michelle knew that in order to hold the trust and loving relations with the gardeners, she had to be accountable to them, above her research agenda.

As she began to bring undergraduate students to the garden, she would explain to them the importance of recognizing their positionality as undergraduates entering a community space.

> If there is something that I usually tell students, especially Latinx students that are coming from spaces similar to Watsonville who are also going through a journey of understanding their positionality and understanding all of their biases and privileges, it is really about understanding where the balance is. You can't separate yourself from the university and that is okay. You gotta just not be weird about it. It is about being raw with people and being honest.

She would share with the students that they should feel comfortable entering into the space and try to overcome being shy and stand-offish. In essence, she was teaching the undergraduates how to be *acomedidos*, to approach the gardeners, say hello, learn what they were doing, and then pitch in. From her perspective, knowing that they were not from the space shouldn't inhibit their efforts to connect; yet connections should happen through showing up, listening and pitching in. It was then that students could then learn alongside the gardeners.

Michelle conducted interviews with gardeners and had to really wrestle with her role as a researcher in the space. She would ask herself, "What are these interviews for? Who will have access to them? What are they going toward?" Then, she would explain to them how her research was part of her work at the university and explain "what the university is doing and how it is playing a role in the space." She continues, "It is important to be really clear about this, but also in personal life. I felt like most of the time, the strongest relationships I have in the garden came from sharing that was not in interviews and reflected in my data, but the raw sharing and conversing with people."

Michelle positioned her role as a researcher as always secondary to her role as a participant in the community garden. Her relationships she built and the conversations she had in the context of the garden often did not make it into her field notes or interviews. Her choice to position her role as researcher as secondary and exclude intimacies from her research was a way in which she cared for the gardeners. Not everything needs to be shared in research. Not all stories are for a research audience. She held the *threshold* of not converting some stories into data (Tuck et al., 2014). As she deepened her connection with the gardeners, she and the gardeners shared intimate stories of migration and state-sanctioned violence against immigrants. These stories never made it into her data. It was a way she cared for her loving relations and also developed herself as an ethically-oriented researcher intent on trying to find ways to be a social science researcher that did not dispossess others (Wilson, 2020) or reproduce the voyeuristic tendencies of social science research (Tuck & Guishard, 2013). It was through these intimate moments, sharing with the gardeners that Michelle felt as if she was humanized. She explains,

> Once they see you for who you really are and not just the academic part of it and not being afraid of being raw and sharing your life experiences... That humanizes you and it builds this bond. I feel the relationships get stronger when you can relate to people in different ways. For me, that was a turning point with the gardeners. After that, you see people differently. There is a magic in being raw with people and being completely honest and not being afraid to share your own experiences.

Instead of succumbing to the urgency of her role as a researcher, Michelle intentionally created time to be with the gardeners, to share and connect. Creating this kind of time and space to be together is a radical act and necessary for building new worlds (Lugones, 2003; Beckett et al., 2012; Beckett, 2020). Through her time in the garden, Michelle healed parts of her own social fragmentation and feelings of alienation. She found a kind of home in the garden and found her own ways back to parts of her humanity that had been fragmented through the violences and dispossession of forced migration. She found connection with a community that reflected various aspects of her immigrant identity; she was able to let her mother tongue flow in a way that brought her back home, not to San Jose, but to Mexico, to her grandfather's farm. And at a time when she felt compounding levels of alienation as a first-generation, woman of color university student, community members in the garden supported her pursuit of higher education, giving her the emotional support and validation she needed to make it through the hardest moments as a university student.

The garden was more than just a place, it was (and continues to be) a pedagogical space where Michelle remembers, returns, revises and reimagines who she is. She has learned to theorize herself in that space and generate ways of being in the world that she then puts into practice. This praxis sutured parts of Michelle back together as a first-generation Mexican immigrant with Purépecha ancestry. I would argue that this process of returning to herself, of connecting with herself and critically reflecting upon her role in the garden was a kind of consciousness raising for Michelle. According to Michelle, it was through both her study in the (H)ACER program and her critical engagement in the garden that she found her way to these understandings and these ways of being. In other words, her study through (H)ACER and the garden provided her the critical and reflexive space where she could become a university student on her own terms (Beckett et al., 2022).

Conclusion

What are the implications of Michelle's praxis for educators interested in creating university programming to support life-affirming community-university collaborations? It is important to note, when exploring this question, that (H)ACER did not create these outcomes for Michelle. As Michelle explains, "You cannot really make those spaces, they can happen. What I think (H)ACER is providing is the space for that to happen, but it is up to people to make those connections, to reach out. You cannot force those relationships. It is really up to the people to make those connections."

(H)ACER provides students with the theoretical and methodological tools and the experiential learning opportunities necessary to embark upon their own journey toward ethically-oriented community-engagement. Guided by popular education and the image of the rocking chair at the Highlander Research and Educational Center (see Chang & Glass, 2022), which symbolizes the radical possibilities of people gathering and sharing about their hopes and dreams, the barriers to those hopes and dreams, and imagining another world is possible, (H)ACER creates space for students to listen to and learn about themselves, each other and the world. At the same time (H)ACER seeks to cultivate this space, radical possibilities of this deep listening and sharing are limited by the fact that the program is in a university of higher learning. (H)ACER engages these limitations and tensions by including in the curriculum a study of the ongoing histories of violence, oppression and dispossession that created and maintain US universities, with special attention to UCSC, which occupies the unceded territories of the Amah Mutsun Tribal Band.

The apprenticeship aspect of (H)ACER allows for students to learn without the undue pressures to produce research (see Lave & Wenger, 1991). They are able to stay in process of their own discoveries and explore relationally. Similarly, the Calabasas Community Garden is an important place for students and families to relax and be together, to learn from each other in non-hierarchical ways, yet there is no timeline to this learning. In other words, (H)ACER provides the space for students to explore their own guiding principles of community engagement and through ethnographic sensibilities put those principles into practice.

Michelle's praxis of loving relations is her theorizing of her experiences and journey in the Calabasas Community Garden that is unique to her, however her practices of loving relations provide astute and important contributions to dwelling in the complexities and challenges of community engaged research and are mirrored in critical scholarship in social science research. Using Michelle's story as a guide, she provokes an important question: How can we, as educators, build community-engaged programming at universities that integrates critical reflexive spaces for undergraduates, learning spaces for community and undergraduate connection, and partnerships grounded in an ethic of care? Addressing this question is at the heart of fostering critical campus sustainabilities.

References

Alexander, J. M. (2005). *Pedagogies of crossing: Meditations on feminism, sexual politics, memory, and the sacred (perverse Modernities)*. Duke University Press.

Beckett, L. K. (2020). Tantear practices in popular education: Reaching for each other in the dark. *Frontiers: A Journal of Women Studies, 41*(1), 120–140.

Beckett, L., Glass, R. D., & Moreno, A. P. (2012). A pedagogy of community building: Re-imagining parent involvement and community organizing in popular education efforts. *Association of Mexican American Educators Journal, 6*(1).

Beckett, L. K., Lu, F., & Sabati, S. (2022). Beyond inclusion: Cultivating a critical sense of belonging through community-engaged research. *Social Sciences, 11*(3), 132. https://doi.org/10.3390/socsci11030132

Chang, E., & Glass, R. D. (2022). Toward a just leadership learning ecology: A CHAT-IT analysis of the highlander idea. *Educational Administration Quarterly, 58*(3), 419–449. https://doi.org/10.1177/0013161X2110548

Engeström, Y. (2001). Expansive learning at work: Toward an activity theoretical reconceptualization. *Journal of Education and Work, 14*(1), 133–156. https://doi.org/10.1080/13639080020028747

Freire, P. (2000). *Pedagogy of freedom: Ethics, democracy, and civic courage*. Rowman & Littlefield Publishers.

Freire, P. (2021). *Pedagogy of hope: Reliving pedagogy of the oppressed*. Bloomsbury Publishing.

Gaztambide-Fernández, R. A. (2012). Decolonization and the pedagogy of solidarity. *Decolonization: Indigeneity, Education & Society, 1*(1).

González, N., Moll, L. C., & Amanti, C. (Eds.). (2006). *Funds of knowledge: Theorizing practices in households, communities, and classrooms*. Routledge.

Gordon da Cruz, C. (2017). Critical community-engaged scholarship: Communities and universities striving for racial justice. *Peabody Journal of Education, 92*(3), 363–384.

Harney, S., & Moten, F. (2013). The undercommons: Fugitive planning and black study. *Research Collection Lee Kong Chian School of Business* (pp. 1–165). Retrieved on January 24, 2023, from https://ink.library.smu.edu.sg/lkcsb_research/5025

la paperson, L. (2017). *A third university is possible*. University of Minnesota Press.

Lave, J., & Wenger, E. (1991). *Situated learning: Legitimate peripheral participation*. Cambridge University Press.

Lugones, M. (2003). *Pilgrimages/peregrinajes: Theorizing coalition against multiple oppressions*. Rowman & Littlefield Publishers.

Lugones, M. (2007). Heterosexualism and the colonial/modern gender system. *Hypatia, 22*(1), 186–219.

Moraga, C., & Anzaldúa, G. (Eds.) (2022). This bridge called my back: Writings by radical women of color. State University of New York Press.

Quijano, A. (2000). Coloniality of power and Eurocentrism in Latin America. *International Sociology, 15*(2), 215–232.

Smith, L. T. (2021). *Decolonizing methodologies: Research and indigenous peoples*. Bloomsbury Publishing.

Stein, S. (2018). Confronting the racial-colonial foundations of US higher education. *Journal for the Study of Postsecondary and Tertiary Education, 3*, 77–98.

Stein, S. (2020). A colonial history of the higher education present: Rethinking land-grant institutions through processes of accumulation and relations of conquest. *Critical Studies in Education, 2*, 212–228.

Tuck, E., & Guishard, M. (2013). (Un)collapsing ethics: Racialized sciencism, settler coloniality, and an ethical framework of de-colonial participatory action research. In T. M. Kress, C. Malott, & B. J. Porfilio (Eds.), *Challenging status quo retrenchment: New directions in critical qualitative research* (Vol. 3, p. 27). Information Age Publishing.

Tuck, E., Smith, M., Guess, A. M., Benjamin, T., & Jones, B. K. (2014). Geotheorizing Black/land: Contestations and contingent collaborations. *Departures in Critical Qualitative Research, 3*(1), 52–74.

Vygotsky, L. S. (1978). *Mind in society: The development of higher psychological processes*. Harvard University Press.

Wilder, C. S. (2013). *Ebony & Ivy: Race, slavery, and the troubled history of America's universities*. Bloomsbury Press.

Wilson, S. (2020). *Research is ceremony: Indigenous research methods*. Fernwood publishing.

Yosso, T. J. (2016). Whose culture has capital?: A critical race theory discussion of community cultural wealth. In L. Parker & D. Gillborn (Eds.), *Critical race theory in education* (pp. 113–136). Routledge.

Linnea K. Beckett is a scholar of education with expertise in decolonial feminisms, informal learning and community-engaged methodologies. She is currently the Assistant Adjunct Professor and Director of the Apprenticeship in Community Engaged Research or (H)ACER program at University of California, Santa Cruz. She co-designed (H)ACER as an anti-racist, anti-colonial community research and learning program to support specifically first generation, BIPOC, and transfer student success at UCSC. Dr. Beckett runs community-engaged research projects through the (H)ACER program, forging deep community partnerships and supporting undergraduate research opportunities.

Michelle Hernandez Romero is an undocumented community organizer born in Michoacan, Mexico. She is a University of California, Santa Cruz alumna, current academic staff at John R. Lewis College, and a Calabasas Community Garden member. She graduated in 2018 with a B.S in Cognitive Science. Michelle served as a Community Outreach Coordinator for the Apprenticeship in Community Engaged Research for College Nine and John R. Lewis College at UCSC.

9 *Plantando Amor y Cultivando Unidad:* A Story of Building and Sustaining an Immigrant-Led Community Garden in Partnership with a University

Michelle Hernandez Romero, Yolanda Perez, and Linnea K. Beckett

Abstract

One project of the (H)ACER program outlined in the chapter prior is the Calabasas Community Garden (CCG). This chapter illuminates the process by which the garden, and a sense of community unity, came to fruition. The CCG is sustained through a collaboration between the University of California, Santa Cruz (UCSC), Calabasas Elementary School, and Watsonville, CA community members. The uniqueness of this collaboration makes it a multifaceted space in which different community members from all partners come together to promote food justice and community empowerment and care. This chapter highlights the collaborative—university, student, and community—approach to research that centers local knowledge and perspectives to bring about transformative change.

Keywords

Community garden · Care · Latinx families · Immigration · Food security · Food sovereignty · Community building · University-community relationships · Self-governance

Introduction

"After a long day of shoveling and spreading mulch on the paths of the garden, Luna left and I kept gardening for another hour. Looking for a dry place to sit, I found Don Luis's chair next to his plot. I always think of him sitting here. Being hugged by the milpa. Facing the mountains. He reminds me of my grandpa, it makes my heart warm and the nostalgia makes me tear up. I sat in his chair and inhaled the scenery. The drying corn leaves on my back, the red onions and fava beans growing in Leonardo's plot. The dormant apple orchard on the other side of the fence. The beautiful mountains in the background. The birds flying around in the gray sky. It reminds me of my abuelito's garden in Michoacán. There's also a beautiful hummingbird with a metallic pink neck enjoying the fava plants. Jovita told me the other day that hummingbirds are messengers, ancestral souls that visit us when we most need them. I feel so warm inside just sitting here. It reminds me of good things and allows me to reflect on the not so

M. Hernandez Romero (✉) · L. K. Beckett
University of California, Santa Cruz, CA, USA
e-mail: mherna81@ucsc.edu

Y. Perez
Calabasas Community Garden,
Watsonville, CA, USA

F. Lu, E. Murai (eds.), *Critical Campus Sustainabilities*, Sustainable Development Goals Series,
https://doi.org/10.1007/978-3-031-30929-8_9

good ones. I wonder if Don Luis feels the same way when he sits here. He lost his wife about a year ago. I remember how Fernanda, his daughter, said coming here keeps him distracted and motivated. She says it's like medicine to him. "As soon as I walk in the house, I tell him 'Papa arreglese para ir al jardin' and he jumps in excitement and in five minutes he's ready to go." I wonder if he actually comes here to think of her and not to distract himself from her being gone. To reminisce about beautiful moments with her. To heal from his loss and feel her love through the Land just like I do when I miss home."—El Jardín es Medicina. Jan 9, 2019. Field note from a day of gardening.

This excerpt, drawn from a field note that I wrote in January 2019, captures a moment during an adverse time of my life when I reflected upon the ways that being in the garden made me feel loved. Being in the garden allowed me to realize that much of my depression was rooted in homesickness. In the same way the garden is medicine to Luis, through connections to our traditional foods, the Land that nurtures us, and the people that bring us love, it is also medicine to me. When I describe the community garden, I always mention how it is the only place that makes me feel like I am back home in Michoacán, Mexico with all my family. I see my *tíos* and my *tías* in many of the gardeners. I see my *primos*, my *primas*, and myself in our youth. Their hard-working spirit, everyday struggles, and even the *'chisme'* (gossip) culture makes me feel at home. I can't imagine graduating from university without having been a part of the Calabasas Community Garden (CCG).

My name is Michelle, and I am a Mexican woman of Purépecha ancestry. I was raised in my native Land, Uruapan, Michoacán until the age of nine when I immigrated to the United States with my mother, fleeing violence and seeking better economic opportunities. I arrived in San Jose, California, the traditional and unceded territory of the Muwekma Ohlone Tribe. I have the privilege of being a Deferred Action for Childhood Arrivals (DACA) recipient since the age of seventeen, which has allowed me to legally work in the United States. I found the garden in my third year of undergrad at the University of California, Santa Cruz (UCSC) when I applied for a research fellowship in hopes of finding meaningful work that was also financially sustainable and supported my skill development. I applied to Supporting Undergraduates by Promoting Education, Research, Diversity, and Agricultural Resilience (SUPERDAR) when I learned that Professor Flora Lu was participating as a research mentor. Through my college affiliation I learned about Professor Lu who was Provost of College Nine and College Ten (later dedicated as John R. Lewis College) at the time. I was inspired by her research work abroad and learning that she was leading food and environmental justice efforts in our local community was encouraging. I was eager to learn from her.

Through my ethics of community engagement (outlined in the previous chapter), I recognize myself as a border crosser (Villenas, 1996), a researcher of the university, and someone with a background and life story similar to those in the community garden. Being a child immigrant, I didn't know what a sense of belonging meant for me for a long time. Understanding the intersections of my identities transformed my estrangement into a thought process that allowed me to see myself as belonging to more than one place. One of these places is the Calabasas Community Garden (CCG).

I write this chapter with Yolanda Perez (Calabasas Community Garden Manager) and Dr. Linnea Beckett (Director of the (H)ACER program). Although Linnea's voice is not foregrounded, my work with her deeply shaped how I make sense of the garden and was foundational to how I tell my story of the garden. Linnea is an anthropologist of education with expertise in decolonial feminisms and popular education, and brought this expertise to build (H)ACER and support the community-university relationship that now flourishes between the university and garden. Linnea and I worked closely with Yolanda Perez, to design and build the CCG starting in 2016. At the same time we built the garden, we conducted research to better understand how people described the value and function of the community garden. In other words, why they chose to be a part of the space. This research generated an important reflexive space for us to think deeply about our praxis of community engagement, which was grounded in listening to the gardener's visions and desires for this space and doing what we could as agents in the university to support and uplift these desires.

Throughout the chapter, I weave together my story with reflections from Yolanda Perez, to describe community-building that was foundational to the emergence of what is now the CCG. I include Yolanda's perspective through excerpts from a series of conversations that she and I had when we began writing this chapter, which I transcribed and translated from Spanish. This chapter is written in my first-person voice and Yolanda is woven throughout in italics, as if she was speaking into the chapter. In using italics, my only intention is to differentiate our voices throughout the chapter.

My name is Yolanda Perez, I am from San Antonio Xicotenco de Morelos, Puebla. I grew up in a small house with a big garden, tending flowers by my grandmother's side since I was able to walk. I came to this country in 2002 and worked as a farmworker for over 10 years in Watsonville, CA. I am a mother to two daughters, Dani and Yuli, a community leader, and the manager of the Calabasas Community Garden. I got involved in the garden project in 2014 when I first heard this idea during a parent-teacher meeting in Calabasas elementary school. Since, I have helped planting seeds and ideas that have grown into this very special place.

As a resilient mother and soft-spoken community leader whose presence weaves as deftly between her carefully tended flower plot as it does within all of the families who share plots alongside her, I, and many in our community, consider Yolanda to be the heart of the garden. We share our perspectives about the garden, not just as a space where people grow food, but as a community hub, a place for celebrating culture, and a space of care caring for oneself and others. My intention is to describe the garden, our community, and the many ways this special place fosters the organic development of meaningful relationships, which we describe as fundamental to not only the sustainability of the garden but to the strong community-university connection we have developed over the years.

This chapter begins with a walk through the garden to bring our readers to the space in which our community building takes place. Then, Yolanda provides her account of the beginnings of the garden project and describes the gardener families first efforts in planting the seeds which became the foundation of the garden. I then share my and Linnea's introduction to the project and my work under Yolanda's leadership while receiving ethnographic research training and community engagement mentorship from Linnea. We follow with an account of the university/elementary school partnership and how it transformed into a university/community partnership through the development of (H)ACER. The next two sections provide an account of the flourishing of the community garden after its first established year (2018), the different relationships being fostered in the garden, and what these relationships mean for the sustainability of the project. I share my research findings as a student fellow and bring in the voices of different gardeners addressing their reasons for participation in the space. The chapter concludes with a reflection on what it means to sustain a community garden in partnership with a university and ends with Yolanda's hopes for the future of the garden and our partnership.

A Walk Through the Garden

Mira que flor tan bonita
qué bien huele esta flor
no la cortes
quiere aire y sol.
Déjala que se columpie

– y se llene de color

Poem by unknown author, painted on a wooden sign by 4th & 5th grade CES students (2018) during the after-school program garden class as a response to flowers being cut from the stem and thrown all around the ground. This sign welcomes you to the garden and lives in Yolanda's garden plot.

The CCG is located on the unceded traditional territory of the Amah Mutsun Ohlone peoples. It thrives on an acre of land at the back end of Calabasas Elementary School (CES), part of the Pajaro Valley Unified School District (PVUSD) in Watsonville, CA. Watsonville is a racially diverse and culturally rich agricultural town about 20 miles south of the city of Santa Cruz. Although the community faces numerous environmental injustices due to pesticide exposure and labor exploitation in the fields and pressing issues such as rising housing costs, lack of medical services, and food apartheid, the Watsonville community has a strong history of resistance and resilience through labor strikes, art, activism, danza, and other community movements that have greatly influenced the resilient culture of Watsonville. Our gardeners are primarily of Mexican descent and the majority work in the agricultural industry, primarily as farmworkers. Most of them also have a history of stewarding and working with food and the Land (see Tuck et al., 2014)[1] at different stages of their lives. Although the garden is located on CES grounds, we consider them two separate entities.

The garden rests beside an apple orchard and looks up to Mount Madonna, part of the southern Santa Cruz mountain range. Small airplanes fly back and forth from the Watsonville municipal airport, located less than three blocks away. Located adjacent to the back parking lot of CES, CCG is identified by a large sign at the front center of the garden that reads "Jardín Comunitario de Calabasas". Large planter boxes all along the front gate are full of medicinal herbs, flowers, blackberries, and chayote plants. In the summer the very back of the garden is a tall green wall of *tres hermanas* (three sisters)—corn, squash, and beans—all planted communally to be shared together. The garden is home to forty-six 15′ × 15′ plots or 'parcelas' tended by individual families. These plots are at the center of the garden, forming a grid; three main water lines providing water access for every plot. Many families have students attending CES and come from nearby neighborhoods, but some families come from as far as the cities of Santa Cruz and Castroville to care for the garden. Each plot tells a story and reflects the culture and traditions of each family. Most families grow at least one kind of flower and it is common to see different varieties of corn, beans and squash native to various regions of Mexico in the plots. Interspersed among the family plots are fourteen smaller 15′ × 3′ plots tended by CES classrooms during the school year. An array of flowers, strawberries, pumpkins, and other vegetables that CES students plant, tend, and watch grow throughout the academic year. When designing the garden, gardeners and university representatives decided to intentionally place the classroom plots at the center of the garden to encourage family-teacher relationships. Yolanda explains,

The teacher plots are an important and special part of the garden. It is beautiful to see that not only the students come to learn about science in the garden, but teachers are learning about their students and their families. This learning helps them build better bonds with their students and I believe that must support their teaching. On the other hand, parents get to know their kids' teachers and this allows them to support their kids' education in ways that they might not have been able to before. It's a space that allows us to be closer to our kids in ways we didn't feel adequate before. For me these relationships are extremely important and valuable. The teacher plots also help teachers learn about the skills behind growing food. For many of us farmworkers it takes us some time to see the value and importance of our work. Of course, the gardening we do here does not compare to the work we do in the fields, but they are skills that often go unrecognized. In this space, many teachers come without much knowledge about gardening and agriculture but they know how beneficial it is for their students' education. That's inspiring for many of us and we are always willing to help them when they ask for it. Kids get to see their

[1] I capitalize Land in this chapter to indicate a primary relationship rather than the general term 'land' to describe a fixed place (Tuck et al., 2014). Many in our community refer to the Land as a teacher, a mother, and mentor who is taking care of us and from whom we are learning in a reciprocal relationship.

parents teaching their teachers. That's something many of us don't get to do anywhere else.

As Yolanda describes, the organization of the garden foregrounds the knowledges and expertise of the gardeners. They are the teachers in the space. Although they do not take on the traditional teacher role as individuals, through cultivating their space, they create an environment that reflects their knowledges, foodways and expertise. When elementary students and teachers enter the garden, they enter this environment and can see these rich histories and practices reflected in the diverse array of plants and cultivation methods.

Collaborative Beginnings

The idea of creating a community garden arose in 2014 during a parent-teacher meeting during which the conversation emerged about repurposing an abandoned lot at CES. The lot was a safety concern for the school and families who had witnessed illicit activities taking place there. Although there were doubts as to whether there would be sufficient interest from families, and concerns about how the construction and general function of the garden would be sustained, there were a handful of parents who were inspired by the proposal and eager to embark on this journey. One of the parents present and immediately interested in this project was Yolanda, whose two daughters attended CES at the time.

When we heard the idea of turning the abandoned lot to a garden for the first time, I thought of the potential of the space to do and address so many things. I thought of a place where we can spend time with family, a place where we can teach our traditions to our kids, and a place where kids can be curious and try new things. I had just finished reading Semillas (see Fleischman & Pedersen, 1997), *a book that was introduced to me through a book club at CES, which inspired me to be a part of this project. I remember the meeting being dominated by families who spoke English. Most of them were very outspoken about disliking and pushing back on this idea. I only listened and then I decided that if I was inspired, there must be others who were feeling the same.*

After the meeting, there were about five other families that shared their interest with me. All of us, the families who were interested, were very shy and did not see ourselves as leaders. We were all farmworkers. We were used to being seen as quiet people who just listened. But I have always been more of a rebel and I will do everything I can to speak up for something that really matters to me. At the time we were working 12–13 hours a day in the fields and people called us crazy for wanting to do this project that was very laborious. But this project meant more than labor to me. One of the reasons this project mattered to me was because when I got here, to the United States, I noticed that not many of the foods and plants from my pueblo existed here. I missed them. Growing up, my fondest memories are from my grandma teaching me how to take care of flowers and I wanted the opportunity to tend a garden here again. Flowers are a big part of my life.

After realizing that there were a good number of us who were interested in turning the space into a community garden, we were able to speak to some teachers and the principal, Mr. Westfall, who connected us to others who were interested in this project as well. We began having meetings to talk about this with families, teachers, and people from the university [UCSC]. Many people would show up to push back on this idea and sometimes to propose other plans. In these meetings, we began envisioning the design of the garden, receiving support from students and staff from the university. When we finalized a design based on the size of the plot, we were told that we could fit up to 48 families. We asked ourselves, where are we going to find so many people? We were just six families, it felt impossible. The principal came to us once and said, "You can come to our meetings and keep encouraging families and inform them about the project. You can make flyers and we can share them with families during pick-up times." We also made it a goal to outreach to families not able to attend the parent-teacher meetings.

We began hosting work days for interested families. Principal Westfall would inform other families through announcements. Slowly, families and students, mainly high school students, would show up to do community service. I remember students who would say they were interested in learning more and would ask things like, "I want to know what strawberry trees look like," which was hilarious to me. I would think, wow there's so much we can teach students from the way foods grow to how to use garden tools such as shovels.

It was hard to get other families to be interested in this project, not only because it entailed physical labor, but also because it entailed time. Many of us had no time to do the physical labor and the outreach to other families. That first year we had seven families with plots in the garden. We had no water access so we would haul buckets of water from the school's closest spigot all the way to the back, where the garden is located. We continued to show up until we were told the district would be installing water piping for us in the space, an effort by Principal Westfall and Professor Flora Lu.

The installation of water piping in the garden was a slow process that took more than a year. With intentions to address concerns about the sustainability and longevity of the project, during this time Professor Lu began efforts to introduce a more consistent presence from UCSC to support the project under the guidance of families who had been at the forefront and leading the project.

An Undergraduate Internship

I began my relationship with the CCG in 2017 when I was accepted into the SUPERDAR fellowship under the mentorship of Professor Lu, Linnea, and, later, *mi querida maestra* Yolanda. My work under SUPERDAR, and then through (H)ACER contributed to the establishment of the garden and provides additional insight into the history of the garden. This year-long fellowship aimed to provide meaningful research experiences with close mentorship to undergraduate students from underrepresented backgrounds who were interested in doing research in environmental and social sciences (Philpott, 2019). I worked alongside Linnea and Yolanda to continue the development of the community garden while simultaneously beginning training in ethnographic research methods during weekly meetings with both Professor Lu and Linnea.

Establishing a consistent presence in CES and the garden and making sense of my positionality were the first practices to which I was introduced. Having learned about the unethical and detrimental ways research has historically been conducted in community contexts, a grounding in anti-colonial and critical feminist concepts and research methods was inspiring. Although I come from a very similar background and hold identities prevalent in the community of Watsonville, I did not grow up or live in Watsonville. Ensuring that I was mindful, self-aware, and critical of how my presence was perceived and received became part of my everyday practice as I continued learning the rich history and pressing realities of Watsonville and its community. I reflected not only myself but UCSC every time I stepped into the garden. I was accountable, before anything else, to the CCG community that had so warmly welcomed me and contributed so much to my life.[2]

After 5 months of turning under weeds as tall as my shoulders, building sheds and pergolas, and measuring out individual plots, I began working closely with Yolanda as a liaison between UCSC and Calabasas gardeners. I followed Yolanda's lead in outreach and organizing and would coordinate with Linnea to design a space that supported family self-governance, autonomy and knowledge practices. In addition, I undertook community outreach with Yolanda, which consisted of creating and passing out flyers to families before and after school to advertise community workdays in the garden. We also organized presentations to share during parent-teacher meetings and other community events. Successful outreach meant showing up to the school during days and hours when families were present, and

[2]Linnea describes my praxis in the prior chapter.

organizing events when families could show up. Slowly, the community garden began growing in membership.

Yolanda recalls our introduction,

When Linnea and Michelle came in, I felt relieved. When I first met Michelle, I felt she was such a genuine and kind person. She was so willing to learn and she would call me every day to ask me how she could help. I was happy I could finally speak to someone who I didn't struggle to communicate with. Although Flora and I had a strong relationship and always did our best to communicate, language barriers create a disconnection that can lead to a lot of things going uncommunicated or misunderstood. I was able to speak to Michelle and express our ideas and needs and she was able to express that to [Flora and Linnea] and vice versa. She became like a bridge.

I was also very happy to learn both [Linnea and Michelle] were here to work long-term in this project and they were not just passing by. Before they came along, we had students who would come to support the development of the garden project; however, they weren't consistent. When they both showed up, we began having more consistent support and presence in the garden from UCSC students. This helped move the project forward a lot faster and brought it more resources we didn't know we would have access to, like funding and outreach support.

I conducted ethnographic field research in the garden over a 1.5 year period, concentrating on the varied interactions and practices in which gardeners, myself included, were participating in. My research was guided by the question, "What are the gardeners' perceptions of the function and value of the community garden?" We wanted to understand how the garden was being used by the community to learn how we, UCSC, could best support the development of the space and our community. The work we did was so embedded in community building that the research felt almost secondary.

During this time, my experience at UCSC completely changed. The CCG and my mentors nurtured my mental health and wellbeing during a time when I was close to giving up on higher education (please see the prior chapter, which describes some of my academic struggles). My personal experiences as an immigrant woman from Mexico allowed me to build meaningful relationships and connections with different members of our community. I began learning about the significance of university-community relationships that work in synergy towards educational equity, student success and social and environmental justice. This is also where I began to understand the significance of long-lasting relationships and the work it takes to nurture and support each other in genuine ways.

The University/Community Partnership

Started by Professor Lu, UCSC and CES began a partnership in 2012. This collaboration established CES as a space for experiential learning through an environmental justice course offered by the UCSC Latin American and Latino Studies Department. This was also the beginning of a more consistent presence of undergraduate students at CES, many who grew attached to the culture and community of Watsonville and CES, and found different ways to continue to show up. Some of these ways include: tutoring CES students, volunteering for school events, giving guest presentations during college week, and providing student support for CES' outdoor educational space, the Discovery Garden. This presence allowed UCSC staff and students to be a part of the 2014 dialogue towards the repurposing of the abandoned lot and the community garden proposal. In 2016, Professor Lu was awarded a USDA NIFA Hispanic Serving Institution Education Grant, which enabled the creation of a food justice coordinator position and led to the hiring of Linnea to support two SUPERDAR fellows, one of them being myself, to continue efforts at CES (Beckett et al., 2022).

In 2018, the Apprenticeship in Community Engaged Research Program, or (H)ACER, was established under Colleges Nine and Ten, led by Professor Lu as Faculty Chair and Linnea as Program Director. Shortly after my graduation, I was hired as the (H)ACER Community Outreach Coordinator. Beckett et al. (2022) write, "Informed by women of color and decolonial feminist methodologies, (H)ACER supports undergraduates, especially those historically and systematically excluded from higher education, to interrogate what counts as research and think critically about their experiences within the university, their own learning, and lives" (1). Collaborating with CES to support the CCG was one of (H)ACER's main endeavors. I saw my new role as an opportunity to continue to support a space and group of people dear to me, while providing opportunities for more students like myself to undertake community engagement, educational programming and research. Through (H)ACER, the CCG became a site where UCSC faculty and students have built relationships with community members, which in turn makes possible experiential learning and the co-production of knowledge. We strive to de-center dominant social, racial, and epistemological hierarchies. The more that local Latinx families govern and take ownership of the CCG, the more enduring and sustainable it becomes.

The Garden Flourishes

With a lot of support, in 2018 the garden signed up 36 families. I couldn't believe it. Many of the new families would tell me that their children told them about the garden and that they too wanted a plot. Little by little, the garden had to be expanded by adding new plots to each side to accommodate families. Many of us got to the garden and continue to be in the garden because of our children.

As the garden grew, there was a need for more resources, but there was also more support as (H)ACER became more established. Included in my scope of work as the Community Outreach Coordinator was working closely with Yolanda to oversee and coordinate the management of the garden and all UCSC student programming in the space. We began hosting more consistent community meetings that supported garden governance and maintenance. This encouraged more participation from families in community work days and events. The efforts of (H)ACER staff and students, the deepening community governance and participation of CCG gardeners, and the funding support from UCSC organizations seeking to foster food and environmental justice education for undergraduates built a solid foundation that enabled the garden to thrive.

Just as Yolanda envisioned, the garden began serving many purposes and providing multiple benefits. Although the promotion of food security and food sovereignty is at the forefront of all the work that happens in CCG, community gardeners have a multitude of motives to engage and continue in the garden.

For many gardeners, this space does a lot more than provide healthy organic food, which is hard for many of us to afford, especially big families. The garden has become a space for stress relief, relationship building, and more importantly a place to nurture our hearts. As immigrants, it is hard for us to find a sense of ownership and belonging. Many of us come for our kids and food, but we stay for that sense of community and ownership. To be able to say, "That is mine and no one can take it away from me." We decide what to grow and how to grow it. For many of us, gardening and agriculture is a part of our culture and who we are. When we arrive in this country, that's one of the first things we lose, and to be able to come back to it and have access to land to grow in has a powerful impact on our immigrant experience. It becomes a place where we can remember our pueblos and the smells of our traditional foods. We have gardeners who grew up in el campo and we have families who grew up in cities and have never grown food. These families still find a part of themselves here.

It is empowering and motivating to feel that sense of ownership and like you have a say in something. When we first began holding community meetings, we were all very shy. I remember how short the meeting would be because no one really wanted to say anything or felt comfortable expressing their opinions. As time has gone by, our meetings get longer and longer. Everyone has comments and suggestions about the way the space functions and we all make decisions together. Before it was "the garden families"; now it's "the garden family." We are all a big family and we look after each other.

In 2018, for the creation of a short video to showcase the progress of the garden, I asked community members, "What does the garden mean to you?" Seven community members volunteered to audio record their short answers. That same year, I concluded interviews I conducted with community gardeners during my SUPERDAR fellowship. In these interviews I asked gardeners,"Why did you decide to join the CCG?" Answers to these questions fell under three main themes:

1. The garden is a community hub that provides a space for gardeners to release stress and spend time in community with family and friends.

> *I got involved the first year to support myself emotionally and occupy my mind a little or rather clear my mind. Afterwards it became more about my sister. We talk while we garden and it has only brought me closer to her.*
>
> *I got involved in the garden because of my dad. To motivate him because growing food is very important to him and suddenly he lives in a small apartment, where he does not have much space. My mother passed away last year and I saw him getting very depressed. He comes here to get distracted and I can see he is motivated.*
>
> *I come to the garden because I like to plant things to eat and also because I get to see my family here and garden with them.*
>
> *This place is very special to me because you can socialize with many families. We come in the afternoon to work for a little while and, well, feel relaxed. Later when we pick the fruit we are always more happy because it's our harvest and it is organic and is more delicious than the ones you buy [at a store].*
>
> *...it is very nice to be here and to spend the day. You can distract yourself a little and feel very well and at ease with the sun, the air, and everyone.*
>
> *I come to the garden for leisure time. It is a very beautiful place and serves as a distraction for me.*

In these quotes gardeners express in different ways that they go to the garden to '*distraerse*', to distract themselves and to be in relationship and community. For many community members the garden is a space to pass time and it serves as an outlet from their everyday lives. They do this by being in community with their family, their plants, and other garden families. Being able to garden alongside their families also fosters the strengthening of family relationships by shifting away from dynamics that they usually experience at home. The sense of community and culture in the garden play a big role in family participation. Being able to build and nurture their family relationships supports families' interest in remaining involved and finding new ways to be together.

2. The garden is a space to grow organic food free of pesticides and other chemicals while continuing culturally relevant practices of cultivation and Land stewardship.

> *I come to the garden because growing food is a tradition for us back in Oaxaca, Mexico. It's something they taught us, it's how we obtain food to survive, to motivate us to give our best effort in life. It has been a way of living for us. It's mine and I feel very happy when I go there.*
>
> *The garden is my culture, it's my roots, it's part of my growth. Here, I have the opportunity to come back to it and be a part of it as much as it is part of me. The garden is part of my spiritual growth, and it's also for maintaining good health. More than anything it's a part of nature, who we should respect and learn to live with. And more than anything, we get to eat more healthily.*
>
> *When I was a little girl my parents always taught me to care for the Land, that the Land is sacred, that the Land gives us food, that the Land has life, that's why the Land grows all the plants and everything we need.*
>
> *I like to plant here in the garden because we get to eat more vegetables without chemicals. The things you buy in the stores have a lot of unnatural things. And honestly, the vegetables, the little pumpkins, everythings we harvest here has more flavor than the vegetables you buy in stores.*
>
> *We like to garden and get fresh produce without chemicals, and that is very good for one's own health.*

In these quotes, gardeners call upon the value and importance of access to organic produce and land access to be able to cultivate. For many gardeners, cultivating is a part of their identity and a piece of them that they often lose when arriving in the U.S. As farmworkers, gardeners know and experience first-hand the dangers of pesticides. They often speak about the genetic modifications that go into mass production of food and how these processes affect the health benefits, quality, and taste of their traditional foods and medicine. The garden provides a space for our community to return to those practices and gain back that part of themselves. The garden also provides a space to address health concerns in their everyday food consumption and be able to grow the food and medicine that allows them to take care of their families and remember their traditional flavors.

3. The garden is a space of learning and knowledge sharing.

> *It brings me joy to come to the garden because of the outcome that we see. The kids can learn about a lot of things like what it means to eat healthier and how to tend the soil and everything in order to obtain healthy products. I also feel very happy because we have gatherings with many families and the kids are able to familiarize themselves with the rest of us. We also can share things amongst each other and with the kids.*

> *I got involved because of my daughters. We wanted them to learn how to harvest food, how things grow, how you plant them. We wanted them to learn something about where we come from.*

> *More than anything for my children… to get more involved with them and do more activities outside of the house with them in which we can teach them things. Also so we can get distracted and be focused on something. Because staying in the house isn't a good option, if we are inside of the house they just want to be in front of the TV.*

> *We decided because my daughter wanted to… we have a few plants in the house and she likes to learn about them. She also wants to plant everything, like the seed of every fruit she eats.*

In these quotes families talk about the garden as space to connect with their kids and be involved in their education. As Yolanda mentioned, for many parents it is hard to participate in their kids' education for various reasons, including language barriers and work. It is common for schools to offer parent involvement opportunities, however, they are often during school hours which are often during parents' work schedules. The garden provides a space for parents to be involved in their kids' education on their own terms. The garden also provides a space for parents to build better relationships with their children and encourage them to dive out of their comfort and media consumption.

Undergraduate Apprentices in the Garden

We [gardeners] love seeing students in the garden. It's a motivation. You see how enthusiastic they are when they arrive and how surprised they get when they see how food grows. It's beautiful being able to teach them something and also to learn about them, their experiences, and where they come from.

The presence of undergraduate students in the community garden to support community projects, workdays, and events are cherished and endorsed by community gardeners. In a community event that took place in the garden in 2018, when asked about her thoughts on student involvement, a community gardener said, "*When I see [undergraduate] students from the university coming out to support the garden, I am inspired by their commitment. They don't participate in the garden the way we do yet they come to work to help it look beautiful. It inspires me to show up to every community workday we have when I know they will be there.*" Student involvement in the garden is fostered by (H)ACER through student led internships, classes, co-curricular programming, and research.

I see that for many of us gardeners, being around the students has made us more outspoken and less shy. When we get to teach them something about the things we love and are passionate about, it gives us all pride and confidence. Seeing them grow is also something that we cherish. I often hear consejos (advice) and recipes being shared. Students leave with bundles of herbs, fruits, and vegetables.

Community gardeners and students working alongside each other allows for organic community-building through conversation and learning, while students are able to support laborious tasks and other financial needs in the garden. The connections that students and gardeners build encourage students to continue showing up to the space beyond their academic commitments. These connections provide a sense of community and belonging when homesickness begins to take hold.

As time has passed, I have seen an evolution in gardeners and student interactions. At first families are very hesitant and shy to speak to the students that came from the university. They would say, "Well, what am I supposed to say to them? They are not going to understand me." I would try to push them and say, "Yes they will and they want to learn from you." Later on, I see them laughing and sharing stories as they garden together. As we begin to come out of our shells, we begin to get excited to see them. Whenever I tell gardeners that a group of students is coming to the garden they immediately ask me what they can do to support their visit and they always show up with food. They want to make sure the students are well nourished and they get excited to talk to them and hear their stories. It's a growth that I have seen amongst all of our community members. We have grown in all kinds of ways here. These kinds of interactions and gaining confidence is our work and who we are encourages us to get involved and do things we have never done before. Many of our community members are more outspoken and take leadership roles in the school and other places. It has helped us remove this layer of fear to be seen or noticed that many of us as immigrants carry. That fear that keeps us at home, stuck in a cycle. Never really enjoying or living our lives. It's amazing how much being a part of the garden has changed our lives. I personally never imagined I would be doing presentations for university classes and one day be called a community leader.

How Do We Sustain This?

The success of the garden is based on the strength of the relationships between the various entities that are so deeply invested in its continued existence. The sustainability of the garden has been possible through transparency and enduring commitment, all of which are centered in all the interpersonal relationships that make this unique space possible. Gardeners are the primary stewards of the space—their sense of belonging and ownership are the backbone on which everything else relies. The gardener community, CES, and UCSC work in synergy to sustain a space that centers education, community care and empowerment.

As a community/university liaison, I have learned first-hand how spaces like the CCG are bureaucratically difficult to create and sustain. There are processes and regulations in place at the university and school district levels that complicate resource sharing and financial sustainability. They inevitably create tensions and instability that need to be addressed sensitively. However, these kinds of relationships are vital to supporting students, especially those from underrepresented communities. Sustaining a community garden is not easy and efforts to simultaneously maintain a university-community relationship does not make this work any less complex. To be in partnership with a community requires nurturing through consistent communication and intentional support. Beyond financial and maintenance labor requirements, co-creating and nurturing such a unique space entails commitment, transparency, and mutual intentions amongst all people involved. CES staff and administration have been immensely supportive of the garden and our community. Principal Westfall, who is a CES alumni, has been a liaison between the school staff, the garden, and the (H)ACER program. The CES administration's dedication to the education and wellbeing of their students goes beyond their daily packed schedules and immensely contributes to the existence of the space.

Yolanda sums it up when she writes,

Thinking about the garden as it is now, I think about how it has gone beyond our imagination and vision of the space since it first began. The garden and its future has always and will always depend on the wellbeing of our community and strength of our relationships. Our community is not just our gardener family, it is also the staff and students from the UCSC and CES. The unity that we have built has to continue to be something that we all want because it is our shared love for our community that has made the garden possible. As long as we all can continue to hope and imagine a future in which we are here and we are together, I know we will be and our garden will continue to be as healthy and abundant as ever. As our community continues to grow, we find ourselves harvesting new skills and understandings of each other and how each of us plays a role in the garden and how the garden plays a role in us.

References

Beckett, L. K., Lu, F., & Sabati, S. (2022). Beyond inclusion: Cultivating a critical sense of belonging through community-engaged research. *Social Sciences, 11*(3), 132.

Fleischman, P., & Pedersen, J. (1997). *Seedfolks*. HarperCollins.

Philpott, S. (2019). SUPERDAR – Supporting undergraduates by promoting education, research, diversity and agricultural resilience. In *USDA research, education & economics information system*. Retrieved January 25, 2023, from https://portal.nifa.usda.gov/web/crisprojectpages/1009021-superdar%2D%2Dsupporting-undergraduates-by-promoting-education-research-diversity-and-agricultural-resilience.html

Tuck, E., McKenzie, M., & McCoy, K. (2014). Land education: Indigenous, post-colonial, and decolonizing perspectives on place and environmental education research. *Environmental Education Research, 20*(1), 1–23.

Villenas, S. (1996). The colonizer/colonized Chicana ethnographer: Identity, marginalization, and co-optation in the field. *Harvard Educational Review, 66*(4), 711–732.

Michelle Hernandez Romero is an undocumented community organizer born in Michoacan, Mexico. She is a University of California, Santa Cruz alumna, current academic staff under John R. Lewis College, and a Calabasas Community Garden member. She graduated in 2018 with a B.S in Cognitive Science. Michelle served as a Community Outreach Coordinator for the Apprenticeship in Community Engaged Research for Colleges Nine and Ten at UCSC.

Yolanda Perez is the Calabasas Community Garden manager and a local community organizer. She was born in Puebla, Mexico and has been living in Watsonville, CA for over 15 years. Yolanda has been at the forefront of the community garden since 2014 and much of her work in the community focuses on food justice.

Linnea K. Beckett is an anthropologist of education with expertise in decolonial feminisms, informal learning and community-engaged methodologies. She is currently the Assistant Adjunct Professor and Director of the Apprenticeship in Community Engaged Research or (H)ACER program at University of California, Santa Cruz. She co-designed (H)ACER as an anti-racist, anti-colonial community research and learning program to support specifically first generation, BIPOC, and transfer student success at UCSC. Dr. Beckett runs community-engaged research projects through the (H)ACER program, forging deep community partnerships and supporting undergraduate research opportunities.

10 Environmental Justice Youth Leadership in Salinas Valley, CA

Karen Crespo Triveño and Agustin Angel Bernabe

Abstract

This chapter showcases Leaders for Environmental Activism Reclaiming Their Health (Leaders4EARTH)'s Environmental Justice Youth Leadership Academy (EJYLA) within the Salinas Valley, an agricultural region in which low-income migrant farmworker families experience disproportionate levels of environmental harm, from pesticide drift to siting of waste facilities. Not only is the EJYLA a space for students to share their experiences of environmental harm, but they are also a way for dreams, hopes, and desires to be re-imagined around the topic of environmental sustainability. This chapter discusses the development and implementation of the leadership academy and its implications for environmental justice education, community health and well-being, and participatory epistemologies.

Keywords

Critical environmental Justice · Youth leadership · Youth activism · Community health · Agricultural communities · Salinas Valley

Help me, and I'll help you grow
Stop polluting my waters and my air
For me to give you the best of me
I'll give you fresh air and a better environment
But you give me peace and tranquility
So help me live longer to see you grow.
– Andrea Castro Corral, 2021

I Am From
I am from the paper houses,
The walls, the dirt, the spiraled fence
Fields surround me left to right
The air I breathe, a cocktail of chemicals
I am from hands of callus from a long day of work,
Rather comforted by their touch
Paulo, his rough beard, covered in dirt
Chelo, her gentle skin, burned by the sun
Tell the struggles of those before and those to come
I am from the fruit of hopes and dreams
A seed of mustard
The pressure, the perseverance, the prayers
Their voices live in my head
"Echale ganas, y Dios dirá"
I am from broken generational barriers,
"La hija, la estudiante, la secretaria"
No longer is my voice buried in the depths of silence,

K. Crespo Triveño (✉)
University of California, Santa Cruz, CA, USA

Leaders4EARTH, Salinas Valley, CA, USA
e-mail: kcrespot@ucsc.edu

A. Angel Bernabe
Leaders4EARTH, Salinas Valley, CA, USA

F. Lu, E. Murai (eds.), *Critical Campus Sustainabilities*, Sustainable Development Goals Series,
https://doi.org/10.1007/978-3-031-30929-8_10

Nor will my sister's
I am from the healing of suppressed words, thoughts, and aspirations
– Magaly Santos, 2021

Introduction

These two poems were written by two Gonzales High School students who participated in multiple Environmental Justice Youth Leadership Academies hosted by the grassroots organization, Leaders for Environmental Activism Reclaiming Their Health (Leaders4EARTH). An Environmental Justice Youth Leadership Academy (EJYLA) is a multi-week, experiential learning program where community organizers facilitate sessions focused on environmental justice, social justice, community well-being, and leadership development themes. The EJYLA encourages youth — who typically range in age from 13 to 18 and live in areas exposed to environmental hazards — to harness their unique strengths to enact positive change in their local communities and beyond. Andrea and Magaly's home is in the Salinas Valley, considered the "Salad Bowl of the World" for its Mediterranean climate ideal for growing broccoli, celery, cauliflower, lettuce, spinach, strawberries, tomatoes, and wine grapes (Askew et al., 2020). This agricultural abundance is made possible through vast economic inequality, labor exploitation, and environmental racism that target migrant farmworkers and children (Flannery, 2017). This agricultural powerhouse exemplifies the political, social, and economic tensions of our food system and how the industrialization, capitalization, and commodification of food have racially altered our relationships to land, food, and place (Wittman et al., 2011). Children who live on the frontlines of large-scale agricultural fields are more likely to experience health issues from pesticide drift and water contamination (Brulle & Pellow, 2006). EJYLA participants often have lived experiences of these injustices. The program supports them in framing and situating their embodied knowledge and personal histories within historical processes of settler colonialism, racial capitalism, patriarchy, and immigration.

Through a deeper understanding of such multifaceted and complex concepts, the youth can re-envision what constitutes an "environmental issue" and what an "environmentalist" can look like. The curriculum of the EJYLA centers on Critical Environmental Justice (CEJ) studies, which addresses the limitations and tensions within prior EJ studies to create a more robust understanding of the root causes of, repercussions of, and forms of resistance against environmental injustice (Pellow, 2016). Youth living on the frontline of environmental hazards should be exposed to conceptual frameworks that relate to socio-ecological threats, settler colonialism, racial capitalism, and mainstream notions of environmentalism. These exposures can support the co-creation of radical imaginaries, as seen in Pulido's approach to making connections between the history of settler colonialism, ecological justice, and spatial imaginaries (Pulido & De Lara, 2018).

Theoretical Frameworks

Critical Environmental Justice

Early environmental justice scholarship has primarily focused on identifying which racial and socioeconomic groups are disproportionally burdened by exposure to toxic pollution (Holifield et al., 2009), relying on methods that looked at spatial correlations of demographics and sites of environmental degradation. However, the causes of environmental injustice are far more complex than what census demographics can show, meaning that distributive outcomes cannot be neatly mapped due to the lack of understanding of different dimensions of justice, such as responsibility, recognition, and participation (Walker & Bulkeley, 2006). Other early scholars argued that environmental injustices can be resolved if the enforcement of health, housing, land-use, and civil rights law could function without discrimination, but only if there is systemic and institutional change through the support of concerned individuals, groups, and organizations (Bullard et al., 2008). Building from early EJ scholarship, the second generation of EJ studies has extended interdisciplinary methodologies to focus deeper attention on other social categories, such as gender, sexuality, and relationships between humans and non-humans (Stein, 2004).

CEJ studies expand the foundational scholarly work from the early stages of EJ studies. Pellow's CEJ framework centers on: (1) the attention on how different social categories are disparately affected by environmental injustices and how they respond to such forms of injustices; (2) the comprehension of how multi-scalar spatial and temporal processes intersect with EJ struggles; (3) the call for transformative approaches towards environmental justice that move beyond the human, state, and capital; and (4) a focus on how all species are indispensable to our present moment and contribute to the building of a just collective future (Pellow, 2018). CEJ grapples with the intersections of social indicators such as identity, socioeconomic status, gender, and sexuality, concurrently with the multiple forms of environmental imjustice and their causes, consequences, and possible remedies (Pellow, 2016). Moreover, environmental justice is a form of praxis, which brings together social movements, public policy, and academic research (Sze & London, 2008). Decolonial CEJ theorists furthermore push back on the assumed universality of western EJ theoretical frameworks, which may perpetuate "victim-centered" justice (Álvarez & Coolsaet, 2020). They call for place-based perspectives to confront the impacts of capitalism, colonialism, and patriarchy and strive towards efforts to achieve self-recognition, autonomy and self-determination (Álvarez & Coolsaet, 2020). New generations of youth activists (Sze & London, 2008) are in the vanguard of critical and decolonial approaches to environmental justice.

Critical Youth Activism Within the Environmental Justice Movement

In rural agricultural regions, the children of farmworker families, often migrants from Latin America, have shown to be subject to pesticide exposure and high rates of childhood asthma (Schwartz et al., 2015). A study by the Center for Health Assessment of Mothers and Children of Salinas (CHAMACOS) found that environmental hazard exposures impact children's neurodevelopmental growth and respiratory systems (Eskenazi et al., 2003). In the case of pesticide spraying, children are disproportionately vulnerable to the effects of exposure, for example, when their parents carry the chemicals home on their work clothes (Goldman et al., 2004). Although there has been extensive documentation on how environmental hazards harm the health of frontline youth, there has been little scholarship on youth activism in the critical environmental justice movement. O'Brien et al. (2018) assert that youth environmental activism research should address different geographic and social contexts and argue that youth will be in a better place to reclaim, reframe, and re-envision their future through critical reflection. Youth participation in collaborative projects can be encouraged by adult facilitators engaging participants in new topics through hands-on activities (Taft, 2015). Hands-on activities can influence youth to engage with the subject, express their ideas, feel a sense of belonging and ask critical questions, which can support them to take charge of their sense of personal and political agency.

Calls for intergenerational justice grow as more youth take charge of their futures, given that adults have shirked their responsibility to sustain a liveable planet for future generations (O'Brien et al., 2018). Youth have found that "adultist models" of youth participation often tokenize them, rather than seeing them as empowered contributors to environmental justice as curriculum designers, facilitators, organizers, or social justice visionaries (Gordon & Taft, 2011). Adults often applaud youth for speaking up about environmental issues, yet their voices are often not taken seriously in environmental decision-making processes (Gordon & Taft, 2011).

Mainstream non-profit organizations in the past have exhibited white saviorism in their work with local communities, without considering the latter's already existing strengths and wisdom that resides within their ancestry, social identities, culture, languages, and historical memory. Youth from historically marginalized communities who confront the burden of environmental racism do not need to be "saved"; they need to be listened to and their expertise recognized. Youth can be the agents of social, political, and economic change to improve their quality of life for themselves and their communities (Schusler et al., 2019). Critical youth empowerment operates through the following dimensions: (1) a welcoming space where youth can be themselves, voice opinions in decision-making, and take ownership; (2) spaces for youth to contribute as they learn new skills; (3) equal power-sharing between youth and adults; (4) opportunities for critical reflection to enhance learning processes; and (5) participation opportunities in sociopolitical processes, where youth are encouraged to participate in action-items that address unjust structures (Schusler et al., 2019).

To create a healthy and just environment for long-term sustainability, education for historically marginalized groups needs to begin at a young age to develop generations of community leaders who can express their expertise if they choose to go into higher education. As community organizers, our vision of an EJYLA is intended to provide leadership development through place-based education (Hilimire et al., 2014), to create accessible environmental and social justice education, and to foster peer-to-peer and near-peer connections. Our EJYLA takes place within communities that live in locations with environmental health threats to build collective youth power through critical reflection, empowerment, and community-based participatory projects where the youth center their lived experiences and embodied knowledge to promote transformative strategies towards social change.

The purpose of this chapter is to showcase how community organizers collaborate with youth living in the Central Coast of California through the programming offered by the EJYLA. The EJYLA program shifts the dominant narrative of top-down hierarchical learning systems and allows every participant to share their lived experiences, identities, cultural backgrounds, and observations from their experiences growing up in the Salinas Valley. Youth must be positioned and taken seriously as the knowledge experts, producers, and stewards for key-decision making processes. Critical environmental justice conceptual and theoretical frameworks must also be understood through theoretical approaches that encompass different angles, social identities, and multi-scalar and temporal processes (Pellow, 2018) that all affect a community's quality of life. In the context of Leaders4EARTH's most recent EJYLA, this manifested as participants explored themes such as settler colonialism and its impact on education systems and food systems, our connection to land and place, community wellbeing, environmental health threats, climate justice, racial justice, mental health, and addressing feelings of "imposter syndrome" as participants created their visions for participatory action research projects.

Background

Environmental Justice Issues in Salinas Valley

According to the Environmental Protection Agency (EPA), environmental justice is defined as the "fair treatment and meaningful involvement of all people regardless of race, color, national origin, or income, with respect to the development, implementation, and enforcement of environmental laws, regulations, and policies." (EPA, 2021). Some EJ scholars have deepened EPA's definition, showcasing that environmental inequality emerges from different forms of inequality at different social categorizations and scales such as race, class, gender, immigration status, nationality, ability, and species (Sze & London, 2008; Pellow, 2018). Indigenous scholar Gilio Whitaker (2019) illuminates how EPA has failed to explicitly name the explicit process of Indigenous dispossession and disenfranchisement in settler societies like the U.S. and fails to integrate Indigenous worldviews, relationships to land, and the more-than-human world. Issues of environmental injustice are prevalent in Salinas, the largest city in California's Monterey County, and a place dubbed "Salad Bowl of the World" in recognition of its agricultural importance (see Fig. 10.1). Stretching 90 miles and less than 10 miles wide, Salinas Valley is a large fluvial valley and the most productive agricultural region (Smith, 2020). A few neighboring cities to Salinas Valley are Gonzales, Chulalar, Soledad, Greenfield, and King City. As many scholars have argued, all these cities have something in common and encounter "real-time concerns for farmworkers such as exposure to toxic substances, occupational hazards, loss of livelihood and cultural identities, intergenerational poverty, injuries and disabilities, chronic stress and disease, involuntary displacement, and death" (Holmes, 2013).

Settler colonialism perpetuates environmental injustices given that it is a "the form of domination that violently disrupts human relationships with the environment" (Whyte, 2018a, b). In Monterey County, the First Nations peoples were the Ohlone/Costanoan, Esselen, and Salinan (Monterey County Historical Society, 2022). Settler colonialism led to three waves of colonization: the Spanish Period (1700–1834), the Mexican Period (1834–1848), and the American Period (1848–Present), all characterized by violence designed to undermine the social resilience, autonomy, and survival of Indigenous peoples (Hoefler et al., 2021). On top of settler colonialism, neoliberal globalization policies from the 1980s introduced free trade agreements such as the North American Free Trade (NAFTA) and

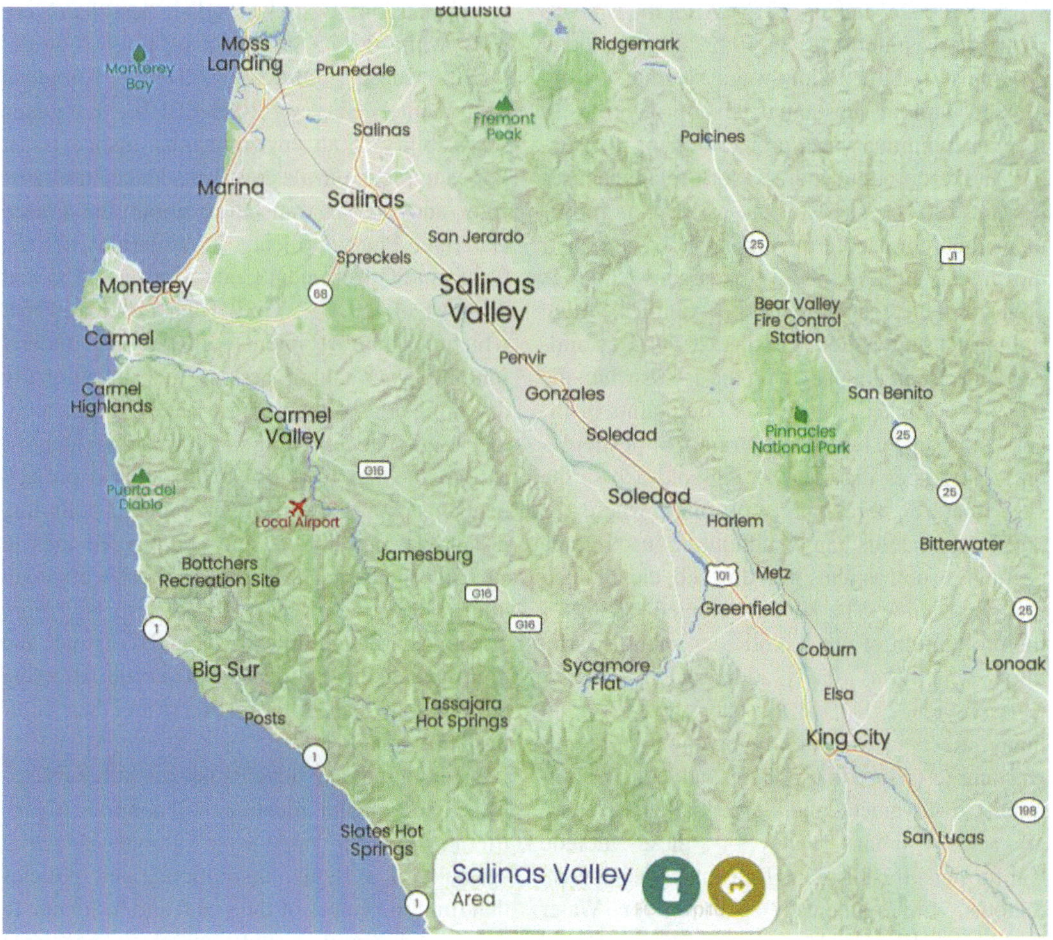

Fig. 10.1 Map of the Salinas Valley. (Mapcarta, 2022)

the U.S.-funded war and violence such as the Central American Civil War, forcing many individuals and families from México and Central America to migrate from their home communities to the U.S. (Hernandez, 2022). Migrants often flee their homes to search for socioeconomic and educational opportunities in the U.S. in an effort to escape fear, human rights violations, climate change, natural disasters, and other environmental factors (United Nations, 2022).

For more than a century (United Farm Workers, 2022), farmworkers have been exploited and discriminated against by large-scale agricultural industries and the government. This led farmers and communities to come together to establish union contracts with the help of key movement organizers such as Ernesto Galarza, Cesar Chavez, and Dolores Huerta (United Farm Workers, 2022; Michals, 2015). After the Bracero program was initiated in 1942, governmental programs recruited low-wage workers from Mexico without substantial labor protections, a federal minimum wage, protections for unionizing, and resources to address work-related abuses (Perkins, 2021). The Chicano civil rights movement incorporated farmworker rights into their movement-building and other issues such as police brutality, educational opportunities, voting rights, and segregation (Perkins, 2021). Key anti-pesticide organizing began in the 1980s when the UFW investigated farmworker complaints of illness and found a high concentration of pesticides on table grapes that were being sold across the U.S. Years of organizing took place where the UFW demanded a ban on commonly used grape pesticides such as phosdrin, dinoseb, captan, and methyl bromide. The UFW continued to call for the prohibition of other pesticides such as parathion, which was finally addressed by the US EPA (Taylor, 2011). To this day, the UFW and many other grassroots organizations are still working to ensure that policies, advocacy, community involvement, and worker rights are equal all across Salinas Valley — these include Californians for Pesticide Reform, Safe Ag Safe Schools, Environmental Coalition for Water Justice (ECWJ), The Agriculture and Land-Based Training Association (ALBA), Community Water Center (CWC), and Asamblea Poder Popular de Gonzales.

Grassroots Organizing in Salinas Valley

In 2011, Roberta Ruiz-Camacho, a community partner and organizer for Asamblea Poder Popular de Gonzales, and the residents of Gonzales filed a civil lawsuit against Salinas Valley Solid Waste Authority for the controversial, polluting Plasco Garbage Plant project (Ruiz-Camacho, 2020). According to the 2010 U.S. census, Gonzales had 8187 residents, of which 88.1% identify as Hispanic or Latino, and 84.3% of Mexican descent. Linguistically, 74.6% of residents speak Spanish at home, and of these individuals, 54.5% spoke English "less than 'very well' while the 25.8% spoke no English at all" (U.S. Census Bureau, 2010, American Community Survey 2006–2010). Language accessibility is an environmental justice issue. For example, Gonzales residents lacked transparency and proper translation about the Plasco Garbage Plant project, a violation of the California Government Code Section 11135 and Title VI of the U.S. Civil Rights Act of 1964, which prohibit recipients of state and federal funding from taking actions that intentionally discriminate and disproportionately impact people based on race, origin, or ethnic group identification. In their response to the Plasco project, residents expressed concerns about air pollution, groundwater pollution, and endangered jobs in agriculture (Urevich, 2013). Their concerns call upon scholars and policy makers to be "more attentive to the multiple power and dynamics and multilayered inequalities and disparities affecting the everyday lives of farmworkers and their families" (Saxton, 2021: 7).

Although the civil rights lawsuit was successful, environmental injustices still continue to proliferate due to large-scale global industrial agricultural systems, institutions, and policies that limit the voices of the youth and residents. In order to dismantle racial oppression, one must use intergenerational methods to train generations of youth EJ activists to create, build, and sustain communities in which all people have equal rights to health, safety, dignity, and a future (Saxton, 2011).

The Environmental Justice Youth Leadership Academy

It is important, before delving into the details about the EJYLA, for us as people and scholars to situate ourselves and where we come from geographically, culturally and intellectually (Davis, 2006; Harrison, 2010; Kovach, 2009; Warren, 2006).

Agustin's Positionality

I migrated from Guerrero, Mexico, in 2001 at the age of six with my mother and two sisters to pursue the "American Dream" due to environmental threats facing my community, such as heatwaves, droughts, clean environment, and food insecurity. We followed the path of many other migrants who have historically rushed to California. A big wave of migrants in California led it to become an agricultural state between 1900 and 1939 (Meade & Guerin-Gonzales, 1996). Although I was thankful enough to arrive in Santa Barbara, I still felt disconnected from all the interactions and people I came across. Not knowing one single word of English, I was frightened to start all over again in a new environment where I was discriminated against based on my race, language, and the color of my skin. Also, being young and uneducated about the different races, cultures, and groups in my area, I had no sense of belonging to a community but my own family. Being raised by a single mother of three, I had to grow up at a young age to help raise my two sisters.

I was thankful to be given the opportunity to get an education and graduate from San Francisco State University, where I received my B.S. degree in Health Education and doubled-minored in Holistic Health and Women's Health. Being a first-generation college graduate came with the hardships that many other migrant youth and families encounter, such as challenges of identity, belonging, stigma, language barriers, and lack of medical care and government support due to the fear of deportation (National Alliances on Mental Illness, 2022). Due to my socioeconomic status, there were few opportunities and governmental assistance. Throughout my journey of learning about my own identity and community, I now identify myself as a Mexican public health and environmental justice activist who is working on de-systematizing the public health systems and racial oppression. I have always envisioned an organization where youth coming from different backgrounds and socioeconomic status can share a space to learn from different perspectives on what environmental health, health and mental health, justice, culture, and self-identity means to them.

Karen's Positionality

I come from Quechua Andean ancestry, and the majority of my family lives in Cochabamba, Bolivia. Like many immigrants, my parents came to this country due to the harsh socioeconomic conditions that their families were enduring. My father's passion for preserving our ancestral music, language, and instruments led him to receive the opportunity to play Andean music in a band and send financial resources back home. After my mother immigrated to meet my father, they decided to reside on Ramaytush Ohlone land to raise my two siblings and me. I grew up in low-income status, often doing poorly in school and working from a young age to help my father sell his CDs as he was street performing in San Francisco. While balancing multiple jobs in high school, I struggled to do well academically and privately battled with my health.

After years of healing, I have been reconnecting with my identity and maternal homeland. I began understanding how global environmental injustices had been targeting my community and homelands for many years. Upon returning from studying abroad in my homelands, I felt fueled to make sense of my ancestral memory, personal history, and the causes of environmental violence, specifically in food, traditional lifeways, and environmental racism. I became the first in my family to have graduated from a four-year academic institution, and I am the first to pursue a doctorate. With this, I hold immense privilege for being a non-disabled, cis-gendered individual with U.S. citizenship currently living on the land of the Uypi tribe of the Awaswas nation.

Early EJYLA Efforts

Our relationship building in Salinas Valley began in 2020, when we were once associated with a San Francisco-based, non-profit organization with funding to hold an environmental and climate justice youth leadership academy in Gonzales. This organization had sent community organizers to facilitate these youth academies for the prior few years, so we were overtaking the work as the newest academy facilitators representing the organization. With no prior teaching experience, we began facilitating the Winter 2020 EJYLA. We developed close connections with Salinas Valley youth through the sharing collective experiences and identities. Most participants were attending Gonzales High School, one was a student at San Francisco State University, and both of the organizers were either recent graduates or current undergraduate students from San Francisco State and University of San Francisco. As time passed, we connected with the youth in numerous ways: the majority of participants were first-generation students with immigrant parents from Latin America; youth were identifying themselves as Chicana/Chicano, Mexican-American, or Latinx; and youth were passionate about developing their leadership skills to protect the health of their community.

We witnessed each other grow through our discomfort of putting ourselves out in the public eye, such as virtual events or community gatherings, where we engaged with political figures such as Monterey County's Agricultural Commissioner. We began working together to create community surveys to ask Gonzales community members if they were aware of the practice, common in rural agricultural communities, of pesticide spraying within a mile radius of people's homes, schools, and community churches. In the proximity of their homes and schools to industrialized agricultural fields, they are exposed to agrochemicals on a regular basis (Schwartz et al., 2015). Children who grow up in the Salinas Valley are exposed to significant levels of organophosphate pesticides due to living near industrial agriculture fields and develop an increased risk of diseases (Stein et al., 2016 & Duramad et al., 2006). Towards the end of this academy, the greatest success that we saw was the fact that all participants felt a greater sense of belonging, that we had created a space where youth felt safe and at home. Ultimately, these cultivating relationships began to blossom as the youth felt inspired to return to future academies to continue speaking up for the rights of their community's health, along with learning new intersections that tie to environmental justice.

As we moved in the Summer of 2020 EJYLA, we were again asked to facilitate a Gonzales youth academy where we were instructed to continue our teaching and develop a pesticide drift community education campaign. We were told by the non-profit organization to exclusively "focus on the environmental justice material" rather than to dive into other connected topics we had previously addressed, such as colonialism, socioeconomic impacts on immigrants working in the agricultural fields, politics associated with industrialized food systems, interpersonal identity, mental health, and overall social determinants of health. We pushed back on the demand, continuing to discuss environmental justice in critical and expansive ways, utilizing interactive lesson plans, open discussions, small-group research activities, games, and self-reflection. We realized that in order to halt the replication of top-down, hierarchical learning structures, we must consciously incorporate opportunities for inquiry on diverse topics that aren't commonly discussed. We incorporated one-on-one mental health check-ins with youth, showing care and recognizing the impacts of learning about oppressive forces such as white supremacy, colonialism, and racial capitalism. We supported youth to find their path in deciding their environmental justice activism projects, reinforcing the power of their agency and action. EJYLA participants undertook social media content creation and dissemination of fact sheets, surveys, social media postings, photography, and videography relating to environmental hazards such as water contamination, pesticide drift, farmworker rights, and mental health.

Toward the end of the Summer 2020 EJYLA, we noticed a shift in the workplace climate of the organization we were working with at the time. After realizing that we were retaliated against for speaking up on our personnel concerns with workplace treatment, we were prompted to leave due to our workplace becoming a hostile work environment. Upon announcing our departure from the organization we were once associated with, the youth were left in shock, confusion and ultimately asked if they could stick with us as we transitioned out of a toxic work environment. The youth emphasized the fact that we needed to stick together as a family of like-minded young people rather than go our separate ways. With optimism, we agreed as we mutually truly felt a sense of hopefulness working with these youth who openly accepted us into their lives and shared pieces of their wisdom. Not only had we made these strong connections with youth from the Salinas Valley, but also with community partners like Roberta Ruiz-Camacho, Salinas Valley Tri Recycles, Safe Ag Safe Schools, and Centro Binacional para el Desarrollo Indígena Oaxaqueño (CBDIO). We built a collaboration with Professor Flora Lu from the Department of Environmental Studies and the People of Color Sustainability Collective (PoCSC) from the University of California, Santa Cruz (UCSC) to continue this work, and founded Leaders for Environmental Activism Reclaiming Their Health (Leaders4EARTH). Since then, we have created and implemented an EJYLA that is closely aligned with our intellectual, personal and organizational commitments, which we implemented in Summer of 2021. This EJYLA centered youth perspectives on environmental sustainability, environmental health impacts, health disparities, socioeconomic disparities, racial injustice, and notions of community health and wellbeing.

Leaders4EARTH's EJYLA

We envisioned our EJYLA as a space where environmental education can be offered for high school students living on the frontline of environmental injustice, while being inclusive for undergraduate students who are interested in CEJ. As we were building the curriculum for our summer 2021 EJYLA, we wanted to build relationships with the undergraduate students who PoCSC was serving. Relationship-building takes time and intentionality, especially during the COVID-19 pandemic which meant that our EJYLA was going to be virtual. To begin this relationship-building process, we collaborated with Christopher Lang and Tashina Vavuris, both ENVS graduate students and key members of PoCSC, to organize three virtual collaborative "Educational Exchanges'" in Winter 2021 to bridge the gap between high school and undergraduate students. We invited 8 Gonzales High School students and 17 UCSC undergraduates to participate in 3 multi-week hour-long exchanges discussing environmental topics and regional challenges of environmental and social justice. These events included interactive presentations incorporating small-group discussions, games, and youth-led facilitation.

After finishing the Winter educational exchanges, we invited the Educational Exchange undergraduate and Gonzales High School participants to plan and facilitate a Salinas Valley Farmers Panel during the Spring of 2021. The Farmers Panel was conceptualized, planned, and facilitated by these students with the support of community organizers, graduate student allies, and UCSC faculty and staff. In preparation for the Farmers Panel, the students recruited panel participants and engaged in recorded, pre-event interviews with the panelists to learn more about their stories, working conditions, and aspirations. The students created the interview questions and facilitated the conversation. Pre-event videos were posted onto Leaders4EARTH social media platforms, so audiences could learn more about the farmers' individual stories before attending the main event. During the main event, youth facilitators held a dialogue with the panelists about their working conditions, experiences with environmental hazards and community health threats, and aspirations toward sustainable farming practices. Community organizers and UCSC undergraduates supported the event with a presentation of environmental justice issues in the Salinas Valley, technology support, live translation, and live comment moderation across multiple streaming platforms. The event was a great learning opportunity for people at UCSC to hear an in-depth and honest dialogue between Salinas Valley youth and farmers. The Farmers Panel inspired a greater momentum building and excitement for Summer 2021 EJYLA, where we invited all the UCSC and Salinas Valley participants from the Winter Educational Exchanges and Spring Farmers Panel.

As exemplified by the Farmers Panel, Leaders4EARTH's EJYLA shines the spotlight on the youth participants and their communities as knowledge producers and positive change agents, who together engage in desire-centered imaginings of a more socially and environmentally just and sustainable world. In the EJYLA, the youth draw from their lived experiences and expertise to design environmental justice activism projects, which entail an outreach plan, bilingual communication with other community members, and utilization of recorded video interviews, fact sheets, social media campaigns, community surveying, storytelling, and other non-conventional methods decided the youth to increase awareness and action addressing an environmental health issue. While the youth saw themselves as emerging leaders, we reminded them that they are also knowledge makers. For example, there are many similarities between activism and ethnographic methods, such as "core methodologies mostly resemble that of grassroots activism: participation, holistic learning, and a humanistic approach to caring, understanding, and working with real people" (Schuller, 2010: 43).

This led to the creation of grassroots working groups with the youth of Gonzales High School to develop youth participatory action research (YPAR). During the Summer EJYLA, we asked participants to create visions of what a participatory action research project would look like to address an issue they've observed in their community. Scholars have argued that participatory action research (PAR) is an important methodology towards social action, which valorizes local knowledge and gives agency to those who are most affected by issues to be the researchers themselves to improve their communities' lives and environments (London, 2007). YPAR expands on PAR to position youth as non-tokenized change agents (Kohfeldt et al., 2011), and as research collaborators for future participation in critical reflection, social action, and empowerment (Langhout, 2019). This helps students get a better understanding and knowledge of their own identity, communities and an opportunity for youth leadership and involvement when developing strategies, policies, and advocacy for their people. The youth from Leaders4EARTH were able to create a series of ethnographic movement methods to not only observe and document, but to react and respond to communities' social, health, and ecological concerns in more mutual, cooperative, and intersectional ways, especially during the COVID-19 pandemic (Besteman, 2015: 260; Saxton, 2015).

To better understand the community and the students, we were able to develop surveys to engage and determine the level of expertise on what the students know about environmental/social justice, environmental racism, community health, and their lived experiences on environmental disparities around their community. There was a total of 10 youth participants ranging from ages 13-to 21 years old. Six youths went to Gonzales High School, while the other four students studied at the University of Santa Cruz (UCSC). The youth participants came from a variety of backgrounds which brought different perspectives on what environmental justice and equity meant to them. Questions consisted of students' backgrounds, the environment, environmental health impacts, notions of community health and wellbeing, and lastly how to improve the Environmental Justice Youth Leadership Academy (EJYLA) to better facilitate and improve Leaders4EARTH as a whole.

When students were asked what their environment means to them, they first thought of "recycle, earth, nature, trees, plants, and the living environment; however, the term can be more generalized and include many different environments when talking in the context of social sciences" (Leaders4EARTH Youth participant 2021). This is true; it's the "web of life" (Graham, 2018), or "where we live, work, play and eat", which may take a toll on their health. Power and inequality shape the lives of communities we work with, and they shape our lives and work as humans. As one EJYLA youth participant said, "having student representatives from different racial/ethnic backgrounds enhances more understanding of how environmental and health disparities are affecting our communities, and people of color, also we are the change and future."

During the 5 weeks of the summer youth leadership academy, we were able to ask students a series of repeated questions to determine their comprehensive knowledge and understanding of the different topics throughout the program. There was a total of 11 questions asked about their connection to their identity, culture, land, environmental justice and health, activism and anything new they learned. One youth participant said, "I learned that everyone has different cultural backgrounds and identities, and not everyone eats the same food on the same holiday." Another youth shared, "this week's topics challenged, confirmed, and deepened my understanding about storytelling, community/public health, and/or the impacts of COVID-19. I found it unique being able to share my input and experiences around COVID-19 while also hearing my classmates' experiences." There are high exposure rates to COVID-19 among socially oppressed groups (Khalatbari-Soltani et al., 2020), as seen with farmworkers and their families who were considered "essential workers" during the pandemic but yet failed to receive government assistance due to their socioeconomic or documentation status.

Week one's themes focused on land acknowledgments, history, identity, and redefining community. We learned from community professionals at the forefront of environmental activism for the respective communities that they identify. We had two guest speakers: Rick Flores, Director of Horticulture and Steward of the Amah Mutsun Relearning Program at UCSC's Arboretum, and Dr. Rebecca Hernandez, then the Director of the UCSC American Indian Resource Center (AIRC). The speakers introduced students to new topics such as settler colonialism, land acknowledgment, history, and redefining our understanding of "community." This gave the youth participants a chance to question and reflect on how the education system and the framing of how Indigenous history is taught in academic institutions. Dr. Hernandez facilitated an activity called "My Food Story," which was a way for us to reconnect with our own respective identities, land, and food. An EJYLA youth participant shared, "it helped deepen my understanding of my cultural identity, environmental justice, and what community building is by having a more clear vision of how history used to be and to have empathy for others."

It is important to have generational education passed down from older activists to younger ones so that culture and history are not lost throughout time. We highlighted the importance of storytelling, community health, social determinants of health, environmental health threats, race, climate change, and environmental justice in weeks two and three. Students could understand how one's socioeconomic status connects with race, health disparities, environmental justice, injustices, community health, environmental health, and current issues like the pandemic. We had the honor of having three guest speakers who are making change within the institutional and community levels: Elida Erickson, Director of the Sustainability Office at UCSC; Aysha Peterson, a Ph.D. student at UCSC whose research is in partnership with ALBA (Agriculture and Land-Based Training Association); and Horacio Amezquita, Director of the San Jerardo Housing Cooperative. One youth participant shared, "Many farmers are at the forefront of climate change and environmental racism. Many governmental bodies at the federal level don't focus on local issues because its heavy work to focus on the issues that happen in each county in the 50 states, so it's important for community members to be active in addressing these issues. Often, community activism is needed." Another student shared, "it's important to have student representatives from different racial/ethnic backgrounds to better understand the needs of their communities, understanding climate injustices, justices and how communities of color are always the most vulnerable and susceptible to the effects of climate change."

For weeks 4 and 5, we focused on themes like praxis (e.g., putting education into action while sustaining community health), mental health, imposter syndrome, and our youth participatory action research projects. Our guest speakers, Dr. Regina Langhout, a Professor from the Psychology Department at UCSC and Sylvane Vaccarino, a Ph.D. student in the Social Psychology Department at UCSC, discussed their activism, YPAR research, and community efforts to our space by applying their expertise to how to maintain health and healing justice within themselves, their work and their communities. This gave the youth the power to use their voices, activism, and community organizing by putting together their YPAR.

The youth participants were divided into two groups during the youth participatory action projects. One group focused on the CHAMACOS Study on Farmworker Women and Children. For this study, the youth participants wanted to incorporate more voices of farmworker women and children by using methods like surveying, storytelling, and interviews to address and bring more awareness of the health concerns and mental health disparities in both women and children brought on by pesticides in Salinas Valley. This was an opportunity for families to share their stories and struggles with one another. The second group focused on the health concerns regarding pesticide drift within their communities. The youth participants' goal was to raise awareness and educate their community about the dangers of using pesticides and pesticide drift. Another interest of the youth was to conduct research and raise awareness about the unfair treatment of farmworkers in Salinas Valley (e.g., how pesticides are sprayed with no warning). The youth participants used their own ways of praxis to develop community-based solutions such as community advocacy, feedback, and strategies for problem-solving.

Conclusion

Community-based solutions/involvement, environmental literacy, health awareness, and education are critical factors in engaging people of all backgrounds to seek new opportunities across multiple scales toward societal, and ecological change. We cannot resign ourselves to the existing forms of oppression, violence, inequities and social-ecological harm (Saxton, 2021), and must continue to protest the systemic inequities that disproportionately affect marginalized groups to heat and pesticide exposure, housing insecurity, and inadequate access to healthcare (Arcury et al., 2002; Flocks & Monaghan, 2003; Horton, 2016; Monaghan, 2011).

Allowing youth and community voices to share their experiences and differences creates spaces like our EJYLA to enhance youth and community activism within their respective communities. Although Leaders4EARTH is in its early stages of organizational development, it is committed to incorporating the voices and needs of the youth and communities of color and building collaborations among different community partners, youth, residents, and stakeholders. It creates a space and opportunity for youth to build their leadership skills, activism, grassroots community organizing, and education to promote healthy and thriving communities within the environmental movement.

Leaders4EARTH's EJYLA can be understood as a co-created space for youth and young adult allies to have courageous conversations on topics that are often not introduced in the K-12 education system, such as environmental racism, immigration, settler colonialism, and racial capitalism. Co-constructed spaces like the EJYLA can support youth in coming to terms with their personal histories, identities, familial migrant histories, and the environmental injustices that exist in their home communities. Their emerging leadership and co-created knowledge in environmental and social justice can support their growth as knowledge producers where they can continuously critique the systemic environmental and social injustices that surround their upbringings while they can simultaneously bring new visions, solutions, epistemologies, and imaginaries throughout their higher education journeys.

References

Álvarez, L., & Coolsaet, B. (2020). Decolonizing environmental Justice studies: A Latin American perspective. *Capitalism Nature Socialism, 31*(2), 50–69. https://doi.org/10.1080/10455752.2018.1558272

Arcury, T., Quandt, S. A., & Russell, G. B. (2002). Pesticide safety among farmworkers: Perceived risk and perceived control as factors reflecting environmental Justice. *Environmental Health Perspectives, 110*(2), 233–240.

Askew, W. R., Adams, M. L., Alejo, L. A., et al. (2020). Monterey county annual crop & livestock report. Retrieved from https://www.co.monterey.ca.us/home/showpublisheddocument/103876

Besteman, C. (2015). On ethnographic love. In R. Sanjek (Ed.), *Mutuality: Anthropology's changing terms of engagement* (pp. 259–284). University of Pennsylvania Press.

Brulle, R. J., & Pellow, D. N. (2006). Environmental Justice: Human health and environmental inequalities. *Annual Review of Public Health, 27*(1), 103–124. https://doi.org/10.1146/annurev.publhealth.27.021405.102124

Bullard, R. D., Mohai, P., Saha, R., & Wright, B. (2008). Toxic wastes and race at twenty: Why race still matters after all of these years. *Environmental Law, 38*(2), 371–412.

Davis, D. A. (2006). Knowledge in the service of a vision. In V. Standford & A. Angel-Ajani (Eds.), *Engaged observer: Anthropology, advocacy, and activism* (pp. 228–238). Rutgers University Press.

Duramad, P., Harley, K., Lipsett, M., Bradman, A., Eskenazi, B., Holland, N. T., & Tager, I. B. (2006). Early environmental exposures and intracellular Th1/Th2 cytokine profiles in 24-month-old children living in an agricultural area. *Environmental Health Perspectives, 114*(12), 1916–1922. https://doi.org/10.1289/ehp.9306

Environmental Protection Agency. (2021). *Environmental justice*. Retrieved January 23, 2023, from https://www.epa.gov/environmentaljustice

Eskenazi, B., Bradman, A., Gladstone, E. A., Jaramillo, S., Birch, K., & Holland, N. (2003). CHAMACOS, a longitudinal birth cohort study: Lessons from the fields. *Journal of Children's Health, 1*(1), 3–27. https://doi.org/10.3109/713610244

Flannery, M. E. (2017). *Schools in farming communities surrounded by dangerous pesticides*. National Education Association. Retrieved January 23, 2023, from https://www.nea.org/advocating-for-change/new-from-nea/schools-farming-communities-surrounded-dangerous-pesticides

Flocks, J., & Monaghan, P. (2003). Collaborative research with farmworkers in environmental justice. *Practicing Anthropology, 25*(1), 6–9.

Gilio-Whitaker, D. (2019). As long as grass grows: the indigenous fight for environmental justice, from colonization to Standing Rock. Boston, Massachusetts, Beacon Press.

Goldman, L., Eskenazi, B., Bradman, A., & Jewell, N. P. (2004). Risk behaviors for pesticide exposure among pregnant women living in farmworker households in Salinas, California. *American Journal of Industrial Medicine, 45*(6), 491–499. https://doi.org/10.1002/ajim.20012

Gordon, H. R., & Taft, J. K. (2011). Rethinking youth political socialization: Teenage activists talk back. *Youth & Society, 43*(4), 1499–1527. https://doi.org/10.1177/0044118X10386087

Graham, W. (2018). *Earth's web of life. Nature's web of life*. Retrieved January 23, 2023, from http://www.freshvista.com/2016/our-web-of-life/

Harrison, F. V. (2010). *Decolonizing anthropology: Moving further toward an anthropology of liberation* (2nd ed.). Association of Black Anthropologists/American Anthropological Association.

Hernandez, J. (2022). *Fresh Banana leaves: Healing indigenous landscapes through indigenous science*. North Atlantic Books.

Hilimire, K., Gillon, S., McLaughlin, B. C., Dowd-Uribe, B., & Monsen, K. L. (2014). Food for thought: Developing curricula for sustainable food systems education programs. *Agroecology and Sustainable Food Systems, 38*(6), 722–743. https://doi.org/10.1080/21683565.2014.881456

Hoefler, A., Reid-Shaw, I., Shaw, D., & Takesy, A. (2021). *Groundwater in Monterey Bay: Challenges and opportunities*. ArcGIS StoryMaps. Retrieved January 23, 2023, from https://storymaps.arcgis.com/stories/7fd09de37536442abb8a4136fe76a896

Holifield, R., Porter, M., & Walker, G. (2009). Introduction to spaces of environmental justice: Frameworks for critical engagement. *Antipode, 41*(4), 591–612.

Holmes, S. M. (2013). *Fresh fruit, broken bodies: Migrant farmworkers in the United States*. University of California Press.

Horton, S. B. (2016). *They leave their kidneys in the fields: Illness, injury, and illegality among U.S. farmworkers*. University of California Press.

Khalatbari-Soltani, S., Cumming, R. G., Delpierre, C., & Kelly-Irving, M. (2020). Importance of collecting data on socioeconomic determinants from the early stage of the covid-19 outbreak onwards. *Journal of Epidemiology and Community Health,*

74(8), jech-2020-214297. https://doi.org/10.1136/jech-2020-214297

Kohfeldt, D., Chhun, L., Grace, S., & Langhout, R. D. (2011). Youth empowerment in context: Exploring tensions in school-based yPAR. *American Journal of Community Psychology, 47*(1–2), 28–45. https://doi.org/10.1007/s10464-010-9376-z

Kovach, M. (2009). *Indigenous methodologies: Characteristics, conversations, contexts*. University of Toronto Press.

Langhout, R. D. (2019). Photovoice as a methodology. In M. B. Schenker, X. Castañeda, & A. Rodriguez-Lainz (Eds.), *Migration and health* (pp. 327–342). University of California Press. https://doi.org/10.1525/9780520958494-018

London, J. K. (2007). Power and pitfalls of youth participation in community-based action research. *Children, Youth, and Environments, 17*(2), 406–432.

Mapcarta. (2022). Retrived from https://mapcarta.com/36590478

Meade, T., & Guerin-Gonzales, C. (1996). Mexican workers and American dreams: Immigration, repatriation, and California Farm Labor, 1900–1939. *Hispanic American Historical Review, 76*(2), 327–328. https://doi.org/10.1215/00182168-76.2.327

Michals, D. (2015). *Dolores Huerta*. National Women's History Museum. Retrieved January 23, 2023, from https://www.womenshistory.org/education-resources/biographies/dolores-huerta

Monaghan, P. (2011). Lessons learned from a community coalition with diverse stakeholders: The partnership for citrus worker health. *Annals of Anthropological Practice, 35*(2), 27–42.

Monterey County Historical Society. (2022). *People and cultures*. Retrieved January 23, 2023, from http://mchsmuseum.com/salinas/local-history/people-and-cultures/

National Alliances on Mental Illness. (2022). *Hispanic/Latinx immigrants and first-generation Americans*. Retrieved January 23, 2023, from https://www.nami.org/Your-Journey/Identity-and-Cultural-Dimensions/Hispanic-Latinx/Hispanic-Latinx-Immigrants-and-First-Generation-Americans

O'Brien, K., Selboe, E., & Hayward, B. M. (2018). Exploring youth activism on climate change: Dutiful, disruptive, and dangerous dissent. *Ecology and Society, 23*(3). https://doi.org/10.5751/ES-10287-230342

Pellow, D. N. (2016). Toward a critical environmental justice: Black lives matter as an environmental justice challenge. *Du Bois Review: Social Science Research on Race, 13*(2), 221–236. https://doi.org/10.1017/S1742058X1600014X

Pellow, D. N. (2018). *What is critical environmental Justice?* Polity Press.

Perkins, T. (2021). The multiple people of color origins of the US environmental justice movement: Social movement spillover and regional racial projects in California. *Environmental Sociology, 7*(2), 147–159. https://doi.org/10.1080/23251042.2020.1848502

Pulido, L., & De Lara, J. (2018). Reimagining 'justice' in environmental justice: Radical ecologies, decolonial thought, and the black radical tradition. *Environment and Planning E: Nature and Space, 1*(1–2), 76–98. https://doi.org/10.1177/2514848618770363

Saxton, D. I. (2011, November 5). Op-Ed: Methyl iodide: Coming to a field near you? *Register-Parajonian*.

Saxton, D. I. (2015). Ethnographic movement methods: Anthropology takes on the pesticides industry. *Journal of Political Ecology, 22*(1), 368–388.

Saxton, D. I. (2021). *The devil's fruit: Farmworkers, health, and environmental justice (medical anthropology)*. Rutgers University Press.

Schuller, M. (2010). From activist to applied anthropologist to anthropologist? On the politics of collaboration. *Practicing Anthropology, 32*(1), 43–47.

Schusler, T., Krings, A., & Hernández, M. (2019). Integrating youth participation and ecosocial work: New possibilities to advance environmental and social justice. *Journal of Community Practice, 27*(3–4), 460–475. https://doi.org/10.1080/10705422.2019.1657537

Schwartz, N. A., von Glascoe, C. A., Torres, V., Ramos, L., & Soria-Delgado, C. (2015). Where they (live, work and) spray: Pesticide exposure, childhood asthma and environmental justice among Mexican-American farmworkers. *Health & Place, 32*, 83–92. https://doi.org/10.1016/j.healthplace.2014.12.016

Smith, A. (2020). *America's salad bowl*. Retrieved January 23, 2023, from https://asmith.ucdavis.edu/news/whither-salinas-valley

Stein, R. (2004). *New perspectives on environmental Justice: Gender, sexuality, and activism*. Rutgers University Press.

Stein, M. M., Hrusch, C. L., Gozdz, J., Igartua, C., Pivniouk, V., Murray, S. E., Ledford, J. G., Marques dos Santos, M., Anderson, R. L., Metwali, N., Neilson, J. W., Maier, R. M., Gilbert, J. A., Holbreich, M., Thorne, P. S., Martinez, F. D., von Mutius, E., Vercelli, D., Ober, C., & Sperling, A. I. (2016). Innate immunity and asthma risk in Amish and Hutterite farm children. *New England Journal of Medicine, 375*(5), 411–421. https://doi.org/10.1056/nejmoa1508749

Sze, J., & London, J. K. (2008). Environmental Justice at the crossroads: Environmental justice at the crossroads. *Sociology Compass, 2*(4), 1331–1354. https://doi.org/10.1111/j.1751-9020.2008.00131.x

Taft, J. K. (2015). 'Adults talk too much': Intergenerational dialogue and power in the Peruvian movement of working children. *Childhood, 22*(4), 460–473. https://doi.org/10.1177/0907568214555148

Taylor, D. E. (2011). Introduction: The evolution of environmental justice activism, research, and scholarship. *Environmental Practice, 13*(4), 280–301. https://doi.org/10.1017/S1466046611000329

U.S. Census Bureau. (2010). Profile of general population and housing characteristics: Social characteristics in the United States for Gonzales City, California. *QuickFacts*. Retrieved January 23, 2023, from https://www.census.gov/quickfacts/fact/table/gonzalescitycalifornia,US/PST045221

United Farm Workers. (2022). *UFW history*. Retrieved January 23, 2023, from https://ufw.org/research/history/ufw-history/

United Nations. (2022). *Migration*. Retrieved January 23, 2023, from https://www.un.org/en/global-issues/migration

Urevich, R. (2013 May 16). Officials to choose vendor for Johnson Canyon project, but activists cry foul. *Monterey County Weekly*. https://www.montereycountyweekly.com/news/local_news/officials-to-choose-vendor-for-johnson-canyon-project-but-activists-cry-foul/article_7106450f-cf68-5381-af4d-629e29f29719.html

Walker, G. P., & Bulkeley, H. (2006). Geographies of environmental justice. *Geoforum, 37*(5), 655–659.

Warren, K. B. (2006). Perils and promises of engaged anthropology: Historical transitions and ethnographic dilemmas. In V. Sanford & A. Angel-Ajani (Eds.), *Engaged observer: Anthropology, advocacy, and activism* (pp. 213–227). Rutgers University Press.

Whyte, K. (2018a). Settler colonialism, ecology, and environmental injustice. *Environment and Society, 9*(1), 125–144. https://doi.org/10.3167/ares.2018.09010

Whyte, K. P. (2018b). Reflections on the purpose of indigenous environmental education. In E. A. McKinley & L. T. Smith (Eds.), *Handbook of indigenous education* (pp. 1–21). Springer. https://doi.org/10.1007/978-981-10-1839-8_66-3

Wittman, H., Desmarais, A., & Wiebe, N. (Eds.). (2011). *Food sovereignty – Reconnecting food, nature, and community*. Pambazuka Press.

Karen Crespo Triveño is a reconnecting Bolivian-American of Quechua heritage. She is a PhD student in the Department of Environmental Studies at University of California, Santa Cruz and one of the co-founders of Leaders4EARTH. As an aspiring scholar-activist, she aims to work alongside transnational social movements across Abya Yala through anti-colonial participatory action research methods. Her research interests include Indigenous food sovereignty, agroecology, seed sovereignty, decolonization, political economy, critical environmental justice, and youth-led activism.

Agustin Angel Bernabe is one of the co-founders of Leaders4EARTH, and was born in Guerrero, Mexico. He is a first-generation graduate from San Francisco State University with a B.S in Health Education and double minor in Holistic Health and Women's Health. He has experience working in the public health sector, community health, environmental health, youth leadership and community engagement.

Conclusion: Sustainability Education and the Future

Christopher Lang, Emily Murai, and Flora Lu

Abstract In this chapter, we reiterate our call for a critical sustainabilities framework that emphasizes the practices of collective care, connection and relationality. Throughout this book, we have given various examples of ways we have engaged in this work in our classrooms, departments, institutions, and communities. We conclude with a focus on the #COLA4ALL wildcat strike that took place at UCSC in late 2019 through early 2020. In addition to its primary call for a cost-of-living adjustment, #COLA4ALL articulated multiple demands to improve the living and working conditions of the university community and to work towards racial justice and long-term economic sustenance in public higher education. The wildcat strike gained momentum and seemed poised to spread to other UC campuses when the COVID-19 pandemic lockdowns occurred. The movement resurfaced in the 2022 labor strikes that were led by four graduate student worker unions across all ten University of California campuses, the largest strike in the history of U.S. higher education. We find hope and inspiration in the ways this movement drew connections and engaged in praxis that clearly articulate the need to advocate for and work towards a broader, more inclusionary approach to sustainability within higher educational institutions.

C. Lang · E. Murai · F. Lu
University of California, Santa Cruz, CA, USA

Keywords Critical sustainabilities • Higher education • COLA • Care • Rent burden • Labor strike • Graduate students

Introduction

Throughout this book, we have focused on unpacking and articulating a critical sustainabilities approach in higher education. We seek to challenge dominant institutional policies, curricula, programming and leadership within higher education that continue to draw from and reinforce normative notions of sustainability and the environment, which privilege techno-scientific, infrastructural and resource management approaches. Many students of color, low-income, first-generation, immigrant, undocumented, and students from other marginalized backgrounds do not see their environmental perspectives or experiences reflected in these approaches, which can perpetuate feelings of unbelonging, exclusion, and disassociation. In contrast, the popularity of the People of Color Sustainability Collective (PoCSC) and the comprehensive data it has collected on student perspectives of the environment show that students from different backgrounds experience, interpret and prioritize environmental concepts differently from each other and from institutional approaches. Because of this, we advocate for understandings of sustainability that are critical, inclusive and just, ones that cener and attend to the needs of the next generation of environmental activists, many of whom are experiencing

F. Lu, E. Murai (eds.), *Critical Campus Sustainabilities*, Sustainable Development Goals Series,
https://doi.org/10.1007/978-3-031-30929-8

devastating environmental injustices in their home communities in real time. They need and deserve to be afforded the tools and knowledge to address what will undoubtedly be ongoing environmental harm. Yet UC Santa Cruz, and other universities, need to support the next generation with meaningful, empowering curricula; relatable mentors, role models and leaders; and co-curricular, experiential, community-engaged and real-world learning opportunities.

In order to go beyond institutionalized forms of mainstream environmentalism, we have advocated for holistic, robust, variegated notions of sustainability that take seriously larger issues of settler colonialism, racial capitalism, and white supremacy. Linking together social justice issues with their historical antecedents enables us to see patterns of large-scale degradation and structural inequality that exist in the present day and manifest themselves in social justice issues that, on their surface, seemingly appear disparate and separate. In the same way that the Blueprint for a Sustainable Campus (discussed in the Introduction) recognizes the interconnections between affordable housing, food insecurity, mental health issues and the sustainability of our environmental communities, or the way the youth activist speakers at the first annual Climate Coalition conference (profiled in the chapter by Lu, Murai, Campbell and Angelo) drew connections between the climate crisis, gun violence, the affordability of mass transit, and environmental health, we believe that for sustainability to be truly meaningful, it must begin to see how contemporary issues originate from troubled histories—centuries-long systems of oppression, genocide and extractivism—that are thus reflected in today's social issues. Working to redress and reconcile these past harms will uncover and illuminate pathways forward that are truly sustainable for all.

For higher education in particular, a holistic approach to sustainability means that we work to address head-on the multiple, deepening inequities within these systems that negatively impact both long-term institutional sustainability and the vitality of surrounding communities. It means that we recognize and attend to the historical underpinnings of our institutions that perpetuate systems of inequality like Indigenous land dispossession; challenge empty rhetorics around diversity, equity and inclusion that act as a cover for institutional protectionism; and protest the growing privatization of our public educational institutions that put economic growth ahead of the well-being of all else: the community, environment, workers. In current times of fiscal austerity and budget cuts, state support for public higher education has significantly dwindled (Mitchell et al., 2018). While more work needs to be done to reverse this trend, universities are reverting to privatized business models that put decision-making into the hands of a few highly-paid administrators who increasingly make non-transparent and centralized, undemocratic decisions behind closed doors. In so doing, they intensify structural and institutional hierarchies; reify extractivist labor politics; reinforce a culture of managerialism and bureaucracy; and continue to respond reactively rather than proactively in the face of multiple crises. Various groups continually find themselves on the short shrift, from faculty who experience the shrinking "shared governance" model, to students whose educations cost more for less, to staff who face inadequate compensation, heavy workloads and few opportunities for advancement, to the rise of the precariously employed who are "flexibly" employed when needed, then casually discarded when not (Washburn, 2008).

It is no surprise that those of us who undertake the work of seeking alternatives to this model and map out meaningful, substantive and sustainable alternatives are the ones most precariously positioned within society and higher education: women, people of color, undergraduate and graduate students, community leaders, youth activists, undocumented immigrants, the tenuously employed. Many of us are on the lower ends of power hierarchies and thus experience the negative impacts of the intersections of power structures most acutely. We are aware that in order to resist pressures to become absorbed into these powerful systems, we must engage with a critical sustainabilities framework that emphasizes the practices of collective care, connection and relationality from within our respective positions and places. Throughout this book, we have given various examples of ways we have engaged in this work in our classrooms, departments, institutions, and communities, from the environmental justice curriculum that empowers environmental justice youth activists in Salinas to the "loving relations" (Lugones, 2003) created with gardeners at the Calabasas Community Garden in Watsonville. These examples show ways that critical sustainabilities can be practiced within different higher educational spaces. Though these examples are highly localized, we believe that they reflect the kinds of cross-institutional community building and future imagining that is possible in different contexts and spaces. It is the collective body of these relatively small acts—these programmatic offerings, orientations and institutional practices in which moments of resistance and human connection are found—that will together create a critical sustainabilities movement. This book is an attempt to recognize, honor and respect our collective efforts in these areas and inspire additional, and differently imagined, efforts in other places and spaces. Throughout the production of this volume, we have attempted to create spaces and relations of care, well-being and solidarity amongst each other in an attempt to infuse an ethic of critical sustainabilities into our own labor.

UCSC on the National Stage: Graduate Students on Strike

The production of this volume was bookended by two influential labor strikes led primarily by graduate students. The first was a wildcat strike originated at UCSC in the Fall of 2019 and crescendoed during Winter of 2020 to virally spread across other UC campuses. Though the strike's primary demand was a cost-of-living adjustment (COLA) for graduate students, BIPOC strikers within this group at UCSC made broader calls to rally all aggrieved stakeholders within the university and hold it accountable, demanding a #COLA4ALL or an Abolish-the-UC ultimatum within its large, diverse base of participants. Nearly three years later, in November of 2022, drawing from the frazzled power of the wildcat strike that COVID-19 subverted, four unions affiliated with the United Auto Workers (UAW) that represent graduate student researchers, academic student employees, post-docs and professional researchers, roughly 48,000 academic workers in total, mounted the largest labor strike in the history of higher education across all ten UC campuses (Lerner, 2023). They were fighting for higher wages, improved benefits, more respect and stabilized working conditions at a time when the California state coffers recorded one of the highest budget surpluses in recent years (Christopher, 2022). Between these events, communities across the world experienced their own relentless streams of discontent, including the COVID-19 global pandemic and its subsequent repercussions such as economic volatility, the isolationism of lockdowns, and a violent, misogynistic backlash against Asian communities; the police murders of George Floyd and Breonna Taylor that re-ignited abolition protests across the United States; assault after assault on American and other Western democracies and on independent, sovereign nations; mass shootings in the most public of spaces—schools, churches, supermarkets—claiming the lives of innocent victims; and devastating climate events, like wildfires, droughts, floods, hurricanes, and sub-arctic freezes. Crisis fatigue in multiple forms has become very real.

We do not see these events as coincidental, but instead as multiple, related symptoms of the failures of the dominant ways we approach our natural world, organize our institutions and social processes, and relate to each other. While we could delve deeper into our individualistic states of cynicism, rage, fatigue and despair, we believe the onslaught of these disasters provides a crucial moment to begin rethinking, retheorizing and reprioritizing how we approach these issues. The time is ripe to imagine alternatives to the world we are living in; indeed, the present day is so fraught with a pervasive sense of helplessness and suffering that it is hard to imagine these times getting any worse. Turning towards one another and developing relations and communities of life-affirming collectivity and care will enable us to seek more productive alternatives and build stronger, more inclusive and just spaces that will sustain us long into the future.

Much of this work of relationality and care was clearly exemplified in the 2020 wildcat strike. Despite punitive institutional threats, various student and community groups came together to work towards a common goal of collective engagement with the university. They resisted the impulses of the university to individuate and separate them in an attempt to dilute their standing and power. By way of conclusion and as a means to reiterate a call for comprehensive, inclusive notions of sustainability within higher education that span traditional boundaries, we turn to an in-depth analysis of 2020 wildcat strike and the work of the #COLA4ALL movement. We find hope and inspiration in the ways this movement drew connections and engaged in praxis that clearly articulate the need to advocate for and work towards a broader, more inclusionary approach to sustainability within higher educational institutions—one that allows all members of the community safe, equitable living and working conditions, while enabling the institution to reproduce itself.

In the remainder of this chapter, Chris Lang, a Black graduate student in Environmental Studies and active participant in the wildcat strike, discusses his experiences of isolation and epistemological dissatisfaction within the Environmental Studies Department and the UCSC campus in his first few years at the university, and the kinds of institutional support he needed to offset his feelings of fatigue and tokenism. Lang contrasts these experiences with the collective nature of the wildcat strike that generated feelings of inclusion, collectivity, and empowerment. While the wildcat strike was halted due the COVID-19 lockdowns and quarantine, the momentum, anger and frustration it generated helped sow the seeds of discontent and renew energy for the much larger 2022 labor strike.

The most recent strike is not without controversy. After nearly six weeks on strike, each union negotiated a new contract with the University of California. In the largest union, the UAW 2865 that represented some 36,000 graduate student teaching assistants and researchers across ten campuses, members were split on ratifying the new contract or continuing to strike. One major issue of contention was the allocation of higher wages for UC's urban campuses (UCLA, Berkeley and UCSF) versus other UC campuses (Santa Barbara, San Diego, Santa Cruz) that nonetheless are located in high cost of living areas. The UCSC bargaining team argued against the contract with 80% of UCSC graduate students rejecting it outright in a UC-wide vote, wanting to continue the strike to push for wages tied to the cost of living, among other concessions. Such a measure would protect members against rising inflation, which leads to an effective pay cut despite higher wages. As 2022 came to an end, the majority of union members voted to ratify the contract at the expense of "rural" campuses like UCSC, effectively ending the strike.

The #COLA4ALL movement is far from over. While the strike achieved a solid labor victory—the four unions won arguably some of the strongest labor contracts within higher education and are inspiring unionization drives, collective bargaining and cross-sector organizing across the country—members of #COLA4ALL continue to agitate for fair living and working conditions. We conclude our book on critical sustainabilities in higher education by focusing on these strikes to demonstrate the importance of collective action to facilitate collective organizing and care; seek alternatives that create inclusive institutions and processes; provide hope and inspiration for sustainable futures; and highlight the ongoing need to engage in multi-sited, multi-faceted struggles for justice.

The "Unrelenting Anti-blackness" of Environmental Studies and UCSC

In 2017, I moved to Santa Cruz after living in Black-centered cities and countries for more than four consecutive years. Returning to California from the Caribbean and Deep South was a cultural shock to my system, like navigating an alternate planet. No more "Hey baby, how you doin?...alright, alright," from my neighbors when I walked down the street past their front door stoops. No more impromptu African drumming sessions under the freeway overpass. I went from being routinely asked "what are you mixed with?" or called "Black creole" or "team light-skin" to being deemed an expert on Blackness in a terrain overwhelmingly devoid of Black people at UCSC and Santa Cruz writ large. To be consistently the whitest person in the room and then consistently the Blackest one in the next is a bit disorienting. While I *am* Black and other Black people in Santa Cruz certainly do exist, the way that Santa Cruz, from its local city politics to university activist efforts, seems very much *in need* of Black leadership and participation, is inescapably tokenizing. Life in Santa Cruz as a graduate student has been expensive, isolated, and emotionally cold in comparison to my previous life of humidity, hospitality, and embedded community.

Noticing the stark void of Blackness in my own department, and the lack of diversity in both faculty and graduate students overall, I joined the diversity committee my first year. I loved the individual faculty members who steered this committee; however, the task seemed too daunting in my opinion to even know where to begin. White space simply cannot be theorized into a diverse, inclusive space. Our efforts would be restricted to re-scripting the department's mission statement, which was a start, but felt performative at best; worse yet, with such lofty language of inclusivity and intentionality, I feared this superficial maneuver, without simultaneously and substantially building out the department's cultural competencies and diversity, would set up incoming minority students for massive disappointment. Overtaxed myself, I could not continue that free work. Fortunately, another BIPOC graduate student and friend of mine picked up the baton. They had the productively disruptive foresight to install an interactive board in our department hallway, titled, how can ENVS be an anti-racist department? The board quickly filled up with colorful sticky notes that offered suggestions or frustrations from graduate and undergraduate students alike: almost all of mine centered on hiring more Black, Latinx, Indigenous, and Asian womnx faculty. I also added that "reverse racism is not real," a response I had expressed in person to a white tenured professor. In class a BIPOC student colleague gave a presentation on environmental gentrification and asked us our thoughts on how one might prevent this worrisome phenomenon from impacting her community back home or elsewhere. When I suggested that maybe the community homeowners could band together collectively and only sell their homes to other BIPOC people, if possible, and to subtly antagonize any gentrifiers who arrive thinking it is a suitable place to establish roots, the professor quickly retorted, "But isn't that reverse racism?"

This professor asking that question is not necessarily bad. I never aim to shame or silence anyone from asking questions. But it shows that within the field of conservation and Environmental Studies, there is a gap in understanding how white supremacy is a historical and ongoing, even unconscious structure or racial project that seeks to protect itself, even as the communities that experience its environmental violence voice their grievances. Many tenured environmental faculty are so focused on protecting "the environment" in its modern disrepair that they fail to consider how colonial, racial policies, like the 1862 Morrill and Homestead Acts and the 1887 Dawes Act that prompted land grant university formation and private property ubiquity, expropriated Indigenous lands and coerced everyone, always unequally, into the ecocidal project of treating land as an ownable, sellable, extractable, and pollutable commodity. These scholars also fail to consider how the widening racial wealth gap ties back to the ecocidal and genocidal project of white people owning enslaved Black people as property. People who were forced to clear cut and transform carbon-sequestering forests and diversely natural grassland prairies into monocropped plantations. People who labored until death to produce wealth for said white families and white-owned businesses as well as the nation for free. Such monumental overlooking of these foundational, eco-social wrecking balls by "the experts" viscerally impacts the quality of environmental education, and the attractiveness of departments like Environmental Studies, for minority students whose communities might experience environmental racism or structural abandonment or a lack of green spaces or an entrenched distrust for white people, yet are nevertheless taught by white faculty who often hold off-putting, unconscious, and quite frankly, ineffective epistemological biases.

Most of my connections with other Black students on this campus initially came from events hosted by the African American Resource and Cultural Center (AARCC), largely due to the great efforts made from its former executive director, Shonté Thomas, and the AARCC interns. Dr. Jackie Rabouin and Mark Gardner were awesome supports in my earliest and mid-years in the program, spearheading Black mental health and intimate community dialoguing spaces on campus. And in my third year I found a lifeline in the Black Geographies Lab (BGL) co-led by professors Camilla Hawthorne and Savannah Shange, and later, Naya Jones. BGL became a community for me to make friendships, receive academic mentorship and career guidance, and connect with other Black scholars at a predominantly white institution in a lily-white city. During and after my department's mandatory course load of my first two and a half years, I intentionally sought to take as many courses as possible that were offered by Black professors, all of whom were outside of my department (there are no Black faculty in Environmental Studies). This habit opened up my mind and scholarly network to consider Blackness and the environment in pretty much any context, from history to sexuality, from education to art. Many of us navigated our respective, isolating academic settings from a spread of departments, supporting each other within and between, on a quarter-by-quarter basis, through an emerging Black grad seminar circuit. We bet on Black. I have so much gratitude for this emergence, which did not exist in my first and second years.

To be Black in Environmental Studies is to be isolated as a person of color. In the six courses where I TA'd Environmental Studies at UCSC (over 600 students total), I can only recall teaching two Black undergraduate students. I remember in a departmental seminar, a white affiliate presented on the great efforts and results his team made in a program that strives to diversify the field of Environmental Studies on campus. The logic went as such:

1. People of color are vastly underrepresented in Environmental Studies/Sciences at this university and beyond;
2. Environmental Studies/Sciences as fields are white-dominant;
3. To address this disparity, we proactively seek to train more BIPOC students to think like environmental scientists.

Through their program, they had successfully brought a noteworthy number of people of color into the field, predominantly Latinx womxn. At the end of his presentation, our department clapped and cheered. I did too, as I was genuinely appreciative of these efforts. Of course, I am in support of diversifying this field! But during the Q&A, I had to bring up the elephant in the room: if environmental science is white-dominant as you say, and we want more BIPOC people to become, or think like, environmental scientists, then aren't we essentially asking BIPOC people to think more like white people, rather than challenging the central tenets of a field that has been historically exclusive in terms of who can hold expertise and what it constitutes as credible knowledge?

Personally, I wonder, what are the environmental implications of doing this type of work (increasing diversity) and not the other type (redefining the field itself, including whose and what sort of work is deemed legitimate and "environmental"), when we look at the overall (in) effective impact of Environmental Studies departments globally to mitigate our contemporary eco-social crises? Having more Black or

BIPOC birders, shark experts, or geologists who replicate white environmental thought and methodology is not going to disrupt ecocidal capitalism that the university was foundational in creating, and is, to this present day, complicit in maintaining and reproducing. I understand that baseline preparation is important in any field to give students a breadth of exposure and training, particularly for students of color coming from low-income and/or urban environments where there may have been less opportunities to engage with "the outdoors." But simply increasing BIPOC enrollment in hegemonically white space is inadequate, for so many reasons, to match the needs of our times. We need to break open the definition of an "environmental scientist" so that Environmental Studies, and future environmental scholars, have sufficient epistemological versatility and depth to dig wide and deep and do the underlying work of excavating both human and non-human habitat destruction or degradation caused by inequality, imperialism, colonization, slavery, consumerism, militarism, toxic wastelanding, and more. When critically analyzed, we notice that all of these issues stem back to hierarchical orders rooted in a triple threat of white, male, and human supremacy.

The institutional bias favoring lucrative STEM research within universities renders any potential remedies to our crises entirely technical rather than social, whether these are climate change ("electrify" the UC but don't give the #landback!) or COVID-19 (lockdown the entire population until a vaccine is ready but don't discuss our preexisting comorbidities from homogenized foodways and ever-present environmental pollutants!) or plastic pollution (invest in chemical recycling solutions but don't meaningfully disrupt this disposable economy) or overfishing (study and map other, more sustainable species' movements to eat but don't stop eating fish!). For more effective environmental and societal crisis prevention and management, universities would be wise to invest heavily in the Arts, Humanities, and Social Sciences, particularly those that uplift Critical Race and Ethnic Studies and Feminist Studies. These departments made up the majority of the lifeblood of the #COLA4ALL constituents, discussed below. Because students of color are less likely to join departments like Environmental Studies, to build out more BIPOC "environmental experts," the university should prioritize hiring diverse faculty in these fields, and not just mainstream environmentalists who might satisfy BIPOC check marks for diversity, but those who prioritize holistic, historically aware, globally Indigenous knowledge production over positivist, profitable, polluting approaches. Many undergraduates of color are coming from communities that experience the burdens of environmental racism, so their participation in developing transformative flight paths beyond our eco-socially catastrophic present is not only paramount, but also equitable and long overdue.

On the Frontlines of the Wildcat Strike

Prioritizing students is especially important in this era where public colleges are now considered part of the "post-welfare state," as higher education has been systematically defunded, privatized, and commoditized (Williams, 2006). In-state tuition at public universities has risen by 175% since 2002, the most of all higher education institutions (Kerr & Wood, 2022). Due to decades of state and federal austerity measures, public institutions now increasingly rely on tuition as collateral to borrow billions of dollars, which can be deployed in ways that lucratively prioritize investments and construction over *instruction* (Meister, n.d.). And as hiking university prices collide with the unrelenting housing market, the costs of higher education and its diminishing promise to guarantee graduates' future financial solvency cause many to ask—like Hasan Minhaj did in his episode of Patriot Act—*is college still worth it?* (Netflix, 2020).

The precarity of graduate students in higher education is becoming well documented (Evans et al., 2018; Flaherty, 2018; Gin et al., 2021). One study conducted by the University of California (UC) concluded that more than one in three graduate students (35%) within the UC experienced depression, and 65% responded that they were "concerned about money lately" (UC Graduate Well Being Survey Report). Testimonials submitted by UCSC graduate students reflected extreme housing and food insecurity. For example, many were only able to secure habitations that were longer commutes away and in worrisome conditions; others responded to the stress of moving incessantly due to landlords pricing them out year after year; some resorted to sleeping in their offices and eating only microwavable and canned foods (COLA Testimonials Digest, 2020). In 2020, graduate students at UCSC made $2434 per month before taxes as teaching assistants, which was paid only nine out of 12 months of the academic year (Karlis, 2020). Graduate students employed by the Ethnic Resource Centers (where I worked the last three years between teaching assistantships), which strive to create community and community programming to support the university's diverse body of students, made roughly $1900 per month before taxes. Meanwhile, the average monthly cost for a one-bedroom apartment in Santa Cruz was $2000 (Zumper, 2020). In this city, over 70% of the renters experience rent burden (Roy & Carlsson, 2021: 49). A special committee on basic needs studied five UC campuses and reported back to the Office of the President that "students prioritized housing over food, and given expensive rent, students had little money left for food" and "students also described coping with a constant tradeoff between eating and other priorities like seeking medical services and physical health" (Special Committee on Basic Needs, 2020). Simply put, as one graduate student representative sent via email to the administrators: *The UC is Making Us Sick, literally.* And to reiterate what many have already exclaimed from the top of their lungs: "TEACHER WORKING CONDITIONS ARE STUDENT LEARNING CONDITIONS!" (McKiernan, 2022). Our financial precarity coupled with increasing work demands trickled down to negatively impact the quality of undergraduates' instruction and feedback. Overworked, underpaid, and backed into a corner of exhaustion and desperation, UCSC graduate students mobilized and voted to engage in a wildcat strike in the fall quarter of 2019, demanding equitable and livable compensation. Graduate students withheld more than 12,000 undergraduate grades from the university as the only source of leverage to negotiate for a cost-of-living-adjustment, which we called a COLA.

UC, UC, you can't hide!
We can see your greedy side!

Early February of 2020. The strike was successful in shutting down the UCSC campus for its third consecutive day. We blockaded roads, protested classrooms, canvassed problematic administrators on bus stop boards. We took over dining halls more than a dozen times, liberating meals for ourselves and countless food insecure students. In protest, we created our own community of learning, an organizing and pedagogical space for theory and direct action and resource distribution and bond-building and information sharing to converge in ways the university was never designed to do. The strike was not only a space to resist the business-as-usual extraction, individuation, competition, and meritocracy that we had endured and internalized; it was where we could reclaim our time, our care, and our collectivity.

Opposing forces confronted one another in the blockaded intersection at UCSC's entrance. Our chants relooped with collective force, cutting through all of the chaotic stimuli to center the topic at hand: what does it mean to work in a system that boasts of its achievements in diversity, research, and service to the community while ruthlessly extracting from and pursuing punitive action against those who make these achievements possible? At the strike site, the moral sides were visible in the landscape: on one side stood justifiably outraged graduate and undergraduate students, many of them students of color, refusing another day of chronic exploitation. *Become ungovernable*, one of the picket-line signs read. *Eat the Regents*, read another. On the other side, 100+ police officers, hailed from Alameda County more than 70 miles away at an exorbitant cost, stood armed in riot gear, ready to use their batons and handcuffs at the behest of campus administrators and their grand puppeteers, the UC Regents. A countdown over a megaphone demanded we vacate the street and move onto the adjacent lawn or face forcible removal and/or arrest. The unruly amoeba of us densified and grounded, bundled up on the hot concrete in collective rings, with our arms interlocked and our torsos facing inwards. UCSC, proudly congratulating itself as "the original authority on questioning authority," had met its match in this seemingly ironic self-fulfilling prophecy (Fly Communications Inc., 2013).

Cops off campus!
COLA in my bank account!
My neck! My back!
The UC is whack!

At the onset of the wildcat strike, a BIPOC cohort quickly formed to expand the frames of exploitation to include multiple campus constituencies. I was invited by a dear friend, another BIPOC grad student, to face the mass of disgruntled but energized graduate students in the amphitheater where we congregated and voted to strike, but we wanted to disrupt the disruption. Instead of a mere COLA, this group demanded a #COLA4ALL, and everyone who had a bone to pick with the university was invited to participate in the mayhem. We articulated the nuanced and overlapping oppressions experienced by many stakeholders within the university, particularly those of color, opening up the aperture of solidarity in a powerful, narrative-changing way that extended far beyond graduate student plights and demands. Pre-existing student activist organizations, such as the Black Student Union, the UndocuCollective, AnakBayan, Worker Student Solidarity Coalition, Youth Democratic Socialists of America chapter, the Snail Movement, and more, could converge in this moment of mass-disruption to call attention to their long-running, unfilled and overlooked demands of the administration. Others, like The People's Coalition, emerged from the latent, fertile soil of protest to create their own ultimata, build unity and sow radical, direct action consciousness among the dissident student crowd. Unsurprisingly, each group's struggles converged at the source of their frustration: UCSC administration. Our efforts interweaved in remarkable collaboration. Our rank-and-file base widened. Our tactics escalated. Some among us were graphic designers and database geniuses who could scrape thousands of emails for mass communication. Others were social media and writing savants, and we had effective public speakers. Some had experience and history leading direct actions; others were spiritually-grounded holders of space. Some became caterer-style chefs; others were sustainable waste management implementers. Just one day of cleaning up and filling dozens of trash bags at the base of UCSC from our inaugural "campus shutdown" was enough to spur adaptation over the weekend. We acquired second hand coolers, boxes, and signage to implement composting, recycling, waste, and water refilling stations to meet the growing waste demands of thousands of striking students. Despite requests for their assistance, we did this without the support of the campus' Sustainability Office.

Our collective included undergraduates, often centering Black students who readily found ourselves out-of-place, underrepresented, overpoliced, and/or increasingly in debt; Indigenous students who make up only half a percent of the entire student body and whose ancestral lands had been violently stolen to found land grant universities such as our very own UCSC; undocumented students whose demands for genuine care (rather than platitudes), a designated office space for interns to work, and more supportive campus environment were ignored, and who experienced chilling retaliation through administration's student conduct proceedings upon taking action into their own hands; lecturers and adjunct professors whose compensation hardly exceeded that of graduate students, with little job security; students with disabilities whose requests for increased accessibility on this mountainous campus were left unanswered; maintenance staff and dining hall workers whose contracts had not been re-negotiated in years despite pleas; graduate students with children who were one paycheck away from homelessness; students who had no choice but to sleep in their cars; queer and trans students who lacked visibility and affirming spaces; and many more. The emotional and intellectual diversity among us was our ultimate strength and collective power. We became ungovernable, removing the psychological and physical cattle guards that separated us by departments, individual achievements, and our specialized knowledges. As one striking graduate student explained, "No matter what threats, no matter what bribes, no matter what anything the university tries to send our way, the strike is spreading. You can't stop an idea" (Solidarity UC Media, 2020).

Relationality and Collectivity in #COLA4ALL: Critical Sustainabilities, Labor Movements, Social Justice

UC, UC, you can't hide!
This strike's going UC-wide!

#COLA4ALL was an autonomous collective by nature, without a direct allegiance or mooring to any branch of the university. Because of this, it was able to move with flexibility and nimbleness. Our leaders were surely identified and targeted, yet the mass of us could quickly fill in the gaps or raise the stakes when facing attempted suppression. From my own training in the Ethnic Resource Centers, I recall learning about the university's organizational chart, which revealed the direct lines of communication and those responsible for entire branches of university functioning. In a similar way that graduate student TA's were the wedge between professors and undergraduates, the strike placed school administrators at a crossroads of allegiance between their one-ups and the students who they vowed to serve. Because the strike emerged somewhat organically, without deference to the university's hierarchical order nor even to the UAW union leadership who initially slandered the wildcat strikers, #COLA4ALL strikers posed formidably untamable and unfazed by typical forms of coercion and strike-breaking. We would not be quelled until the lockdown literally shut everything down.

What the wildcat strike gestured to, even if imperfectly, was a sense of relationality across all stakeholders within the university. Stationed at the field and intersection of the base of campus for over a month, an unpredicted benefit of the strike was a renewed and reclaimed sense of time granted to all the students, graduate and undergraduate participants alike. Freedom was in the air. The strike brought us together, day after day. Many of us lost track of the actual demands, optimistically, if not naively, enveloped by this surreal alternative to the chronic grind. Deep, interlocked, and repeating yet tense encounters created bonds of interdependency, intimacy, and trust that countered almost everything the university upholds, which are values of individual meritocracy, competition, and punctuality. In a normal academic setting, the bathtub of tasks from reading to grading to researching to writing is forever overflowing, so we never have the time to build these relationships and reset our restless minds and bodies. The productivity treadmill keeps us so busy that we are unable to change the unsustainable systems that we collectively suffer from but in isolation. Creating time by congregating and disrupting the ecocidal norm can help us socially transform many of the problems we experience, and many of the environmental impacts that we contribute to through individualism.

The late priest Malidoma Somé (1994) taps into the importance of a relational epistemology when explaining the difference between his learning in seminary school in Burkina Faso and that of his Indigenous Dagara tribe:

> What I was learning [from my tribe] made sense only in terms of relationship. It was not fixed, even when it appeared so. For example trees are not immobile, they travel like us from place to place. By contrast, I could see that the Western knowledge I had been given had the nature of a solid because it is wrapped in logical rhetoric to such a degree that it is stiff and inflexible. The learning that one gets from a book, from the canons of the written tradition, is very different from the living, breathing knowledge that an elder has to offer—and different from the knowledge that comes from within, from the soul (204–205).

Unlike the written Western teachings of an inherent, positivistic truth that is "stiff and inflexible," unchanging once on the page, Somé unpacks a context-specific type of knowledge that exists forever in relationship. The obsession with objectivity in Western knowledge production opposes a porous, Indigenous knowingness that emerges "from within, from the soul."

Somé's life straddling seminary schooling and his tribal existence was a tug of war between two incommensurable ways of learning and knowing. He highlights how the "living, breathing knowledge" of elders recognizes tree agency and mobility. Nishnaabeg scholar Leanne Simpson echoes this, affirming that knowledge is a combination of heart, emotion, thought, and intellect. In her tribe's creation story, "Gzhwe Manidoo transferred all the knowledge that went into the creation of the universe to Nishnaabeg bodies, but that knowledge was so vast it didn't just stay in our heads, it spilled into every aspect of our beings" (Simpson, 2017: 28). Such an education is *alive*, and it "comes from the roots up. It comes from being enveloped by land" (ibid). On the contrary, Western pedagogy, particularly STEM fields within it, treats the student like an empty vessel to receive information through the mind alone (Freire, 1968), instead of considering the entire body and its full range of sensations in relation to its surroundings as integral to one's knowingness.

Undoubtedly, we tapped into a form of living, ancestral, and full-bodied knowledge during the #COLA4ALL strike. Despite the institution's departmentalized design to ensure its graduate students collectively suffer in isolation (and thus, fail to organize for better working conditions), we co-created space to exhale and heal through resistance in affirming, relational, and expansive ways. We eschewed colonial constructs of time and deadlines that were built around the "work day" and maximizing productivity. We made altars, performed ceremonies, hit drums, and shared prayers and chants for ancestral guidance to ground these actions away from the trappings of egoism—particularly as the strike gained more external attention—and towards the goals of bringing accountability, fairness, balance, and liberation for all. The destructive energies people carried were ultimately bound to visions of constructing something entirely different, wholesome and sustainable.

The pandemic lockdown truly saved the UC. Similar to the way the UC lacked transparency with its decision-making, we saw this from the top-down in our own government, which prioritized techno-scientific biomedical expertise without considering any possible social scientific solutions aside from mass isolation. This conveniently siphoned even more of an unprecedentedly unequal society's power and wealth to the very top: in less than two years, billionaires became $4.4 trillion richer, and over 100 million people fell below the poverty line (Stiglitz, 2022). Leading up to the COVID-19 lockdown, the wildcat strike placed UCSC first, and then the entire UC enterprise shortly after, in an existential crisis. The unpredictability was unsettling in the best of ways. Students were calling the shots, teaching one another about radical histories and direct-action strategies in real time with real life consequences. The chilling effect of two years of mandated isolation and striking students' graduation and/or suspension, combined with the assumed re-strategization of leadership within the UC made it nearly impossible to keep that same energetic spirit alive. In the Spring of 2022 at the John R. Lewis College inaugural commemoration and naming ceremony, a Black Student Union-led undergraduate resistance, organized by many of the same students who protested beside us two years before, rekindled this energy. They refused performative, "anti-racist" niceties like college renaming after civil rights icons when this same university sicced armed police on these same students just two years before. When the COVIDian glitch in the universe left their own demands of the university largely untouched, and the university unscathed. Student organizers printed off sheets of paper that read "UC, we can see your greed," which they distributed down the aisles of the Quarry Amphitheatre for audience members to hold up whenever UCSC Chancellor Cynthia Larive and UC President Michael Drake spoke. Subtle. Subversive. They thought we forgot. They thought they were safe. The crowd murmured with curiosity. The entire event was thereafter transformed out of a place of scripted celebration and to one of precarious, passionate, radical truth-telling. Their brave efforts took my mind back in time for a fleeting moment to that place of people power, and I felt my life force, which had been coerced into a two-year slumber, return via the goosebumps on my arm.

I have been conducting field work in New Orleans and unfortunately was not very active nor engaged in the recent UAW strike efforts. Since I left Santa Cruz in summer 2022, it was refreshing to witness a speedy recovery time of resistance from within a year of being back in person, and to be informed of updates by our department's graduate student WhatsApp group chat and other email lists. A friend texted me at some point saying that College Nine and John R. Lewis College dining hall being reclaimed, stoking memories of our pre-COVID-19 resistances that I had nearly forgotten. The writing was on the wall, though: this time, with two years of preparation, UC knew what was coming and knew it could not escape some form of graduate student capitulations. By ensuring a formalized negotiation process take place with status quo union leaders in 2022, after the convenient suppression of effective guerilla tactic disruption by rank-and-file wildcat strikers by COVID-19 in 2020, the UC could ensure its own sustainability. Wildcat strikers' demands for a cost of living adjustment—a radical concept that would have hitched our compensation to the housing market, thereby disincentivizing the UC from setting exorbitant housing rates on campuses—were predictably removed from the contract.

Conclusion

In *Undrowned: Black Feminist Lessons from Marine Mammals*, Alexis Pauline Gumbs (2020) envisions a thought-provoking shift in our way of schooling, one that appropriately captures our aspirations for making higher education truly sustainable. She asks:

> What if school, as we used it on a daily basis, signaled not the name of a process or institution through which we could be indoctrinated, not a structure through which social capital was grasped and policed, but something more organic, like a scale of care. What if school was the scale at which we could care for each other and move together. In my view, at this moment in history, that is really what we need to learn most urgently (55).

Undrowned poetically unpacks the ecocidal and genocidal legacies of slavery, colonialism, taxonomy, techno-science, and commodification writ large that are rooted in centuries of capitalistic pursuits and plunder. The book provides us with a series of rehabilitating offerings inspired by our long-lost marine relatives, salves that gently assist us in dissolving an entrenched human/nature divide, internalized productivity and separation, and unconscious extractive actions, so we can eschew white thought and behavior and all of its failing, unsustainable, unbreathable conditions. Gumbs (2020) writes, "I just wanted to know which whale was which, but I found myself confronted with the colonial, racist, sexist, heteropatriarchalizing capitalist constructs that are trying to kill me—the net I am already caught in, so to speak" (6).

The practices of care and collective movement that Gumbs uplifts are ancestral technologies that we have inherited for our survival, which is our sustainability. They are eco-social orders in and of themselves that have been subverted by the overdetermined, anti-relational intrusions of capitalism. To center care and communal well-being in an educational setting, never just for ourselves but for all of biota and the earth, would require us to scale up our understanding of relations to whom we are accountable. As scholars dedicated to transforming how we learn, how we know, and how we teach about our environmental relations and interdependencies, we see the offerings from *Undrowned* as necessary eco-social interventions for the multi-faceted crises of today. The violences that continue to be committed under the disciplining of environmental education, expertise, and knowledge production, and the ongoing overlooking of ancestrally imperative values that Gumbs uplifts—such as collective care, relationality and ferocious, vulnerable presence—can and must be foregrounded and challenged at the level of institutions.

This surely demands the overhaul of our unsustainable curricular pace, prescriptive learning agendas, canon inundation, and syllabi reading assignments that "prove expertise" yet leave the most important lessons involving community values off the table. Slowing down, if not downright stopping, at the pace of community permits us to connect our external environments with our internal intuitions—doing so, we more holistically understand ourselves, our passions, one another, and our world. We redesign outdated infrastructures to match the needs of the collective, or we dismantle them and salvage the parts. We democratize space with circular learning encounters rather than linear, hierarchically-imposed, sink-or-swim style, predetermined "teaching" models that glorify sleepless nights in libraries for final exams and indoctrinate our most enthusiastic learners into a world of competition. Even in natural science settings, there is less and less time-intensive field immersion in pedagogy and more emphasis on technology-driven data collection, computations, and modeling (Anderson, 2017). Our current "earth-saving" work sector prioritizes people with Environmental Studies and Sciences degrees, yet people being pumped out of this knowledge-production system are often taught to see the natural world as outside of ourselves. Rather than learning the skills to exist in good relations and balance with earth, we are taught that in order to survive, we must first get a job in our related field.

The #COLA4ALL protest of 2020, and the BSU's follow-up demonstration at the John R Lewis inaugural college naming ceremony two years later, reveal how marginalized students of color see through the veil of progressive talk and challenge the university to walk-the-talk. This current system premised on fatigue and overexertion is how we mass-produce a culture of cheating and corner-cutting and technological inventions like ChatGPT: because of the perverted learning process and the severe stakes of failure, youth are overwhelmingly and helplessly coerced to focus on outcome. From the get go, collective movement is fractured into individual exceptionalism, and the environment (read: ourselves) experiences this breakdown from overextraction and pollution due to entrenched notions of private ownership and competition over collaboration. If racial capitalism drives ecocide (it does), and the university is a primary input to the engine of racial capitalism through its student-to-worker pipeline (it is), then a sustainable university is an oxymoron. While genuine efforts and people seeking to green and decolonize the university surely do exist, we have to ask ourselves: is the university more interested in sustaining itself or sustaining the people and the planet, particularly when these priorities appear to be increasingly mutually exclusive? Can the university restore its wrongs to bring us back into a place of wholesome balance? Which people and places on the planet are it hoping to sustain, and which will it sacrifice for its pursuits of knowledge and profit?

Clarke (2022) argues that "schooling does not cultivate the experimental capacities needed to grow and heal one's community amid ecological precarity, insofar as the process of healing is inextricably bound to the overturning of the political economic order sanctioned by the education system" (117). One can not be sustainable and simultaneously mum, or worse yet, explicitly dependent on "an economic system that privileges profit and fabricates racialized, disposable beings" (Verges, 2017). We live in a world that we already know can hold us in our fullness when we exist in right relations between ourselves, our ancestors and relatives (human and more-than-human) across the globe. We must redouble our efforts and do the work that can help us, from diversifying under a common cause and collectivizing, to redistributing power and resources. So when we see through the veil and protest and disobey in masses and then face punishment, retaliation, and isolation, we know we are hitting a nerve, that we have done something right. #COLA4ALL was about so much more than financial compensation. In this modern academic apocalypse where students spend more time on screens, in cubicles, in debt, and in distress, #COLA4ALL articulated that another learning environment that prioritizes relationships over accomplishments is possible, knowing that our community within and beyond the university holds a limitless potential in ourselves to "refuse" and "slow down" and "go deep" as Gumbs envisions, so we can all find a place in return and renewal.

References

Anderson, J. G. (2017). Why ecology needs natural history: The two fields' intertwined histories show that most theoretical breakthroughs are preceded by the kind of deep observational work that has fallen out of vogue in the past half century. *American Scientist, 105*(5), 290–298.

Clarke, A. (2022). Songs of school abolition. *Curriculum Inquiry, 52*(2), 108–128.

Christopher, B. (2022, May 13). *California's $100 billion surplus: What to know about Newsom's spending plan*. CalMatters. https://calmatters.org/politics/2022/05/newsom-budget-spending/

COLA Testimonials Digest. (2020, January 16). https://drive.google.com/file/d/1-2mE47dlYrxKF_Ugw3WehOJftuTlie9s/view

Evans, T. M., Bira, L., Gastelum, J. B., Weiss, L. T., & Vanderford, N. L. (2018). Evidence for a mental health crisis in graduate education. *Nature Biotechnology, 36*(3), 282–284.

Flaherty, C. (2018). New study says graduate students' mental health is a "crisis.". *Inside Higher Ed*. Retrieved January 29, 2023, from https://www.insidehighered.com/news/2018/03/06/new-study-says-graduate-students-mental-health-crisis

Fly Communications Inc. (2013). UCSC the original authority on questioning authority. *Vimeo*. https://vimeo.com/76995944.

Freire, P. (1968). *Pedagogy of the oppressed*. Seabury Press.

Gin, L. E., Wiesenthal, N. J., Ferreira, I., & Cooper, K. M. (2021). PhDepression: Examining how graduate research and teaching affect depression in life sciences PhD students. *CBE—Life Sciences Education, 20*(3), ar41.

Gumbs, A. P. (2020). *Undrowned: Black feminist lessons from marine mammals*. AK Press.

Karlis, N. (2020, March 4). 'Not surprising, but yet unbelievable': UC Santa Cruz students fired over wildcat strike. *Salon.com*. https://www.salon.com/2020/03/04/not-surprising-but-yet-unbelievable-uc-santa-cruz-students-fired-over-wildcat-strike/

Kerr, E. & Wood, S. (2022, September 13). A look at college tuition growth over 20 years. *US News & World Report* https://www.usnews.com/education/best-colleges/paying-for-college/articles/see-20-years-of-tuition-growth-at-national-universities

Lerner, M. (2023). From the Front Lines of the Largest Academic Strike in US History: Graduate Students Across the UC System Go On Strike. Retreived June 27, 2023, from: https://www.versobooks.com/blogs/news/5507-fromthe-front-lines-of-the-largest-academic-strike-in-us-history-graduate-students-across-the-uc-system-go-on-strike

Lugones, M. (2003). Pilgrimages/peregrinajes: Theorizing coalition against multiple oppressions. Rowman & Littlefield Publishers

McKiernan, E. (2022, September 26). *Twitter*. https://twitter.com/EmilieMcKiernan/status/1574491741708861472

Meister, R. (n.d.). *They pledged your tuition*. The Council of UC Faculty Associations. Retrieved January 29, 2023, from https://www.cucfa.org/news/2009_oct11.php.

Mitchell, M., Leachman, M., Masterson, K., & Waxman, S. (2018). *Unkept promises: State cuts to higher education threaten access and equity*. Center for Budget and Policy Priorities. Retrieved January 29, 2023, from https://www.cbpp.org/research/state-budget-and-tax/unkept-promises-state-cuts-to-higher-education-threaten-access-and

Netflix. (2020, June 15). Is college still worth it? | Patriot act with Hasan Minhaj. *YouTube*. https://www.youtube.com/watch?v=YytF2v7Vvw0&t=1s

Office of the Presidents. (2021, January 20). *Update on exploring housing insecurity among University of California Students*. Retrieved January 29, 2023, from https://regents.universityofcalifornia.edu/regmeet/jan20/s3.pdf

Roy, A., & Carlsson, C. (2021). *Counterpoints: A San Francisco Bay area atlas of displacement & resistance*. PM Press.

Simpson, L. B. (2017). *As we have always done: Indigenous freedom through radical resistance*. University of Minnesota Press.

Solidarity UC Media. (2020, March 8). UCSC Graduate students and allies respond to being fired. *Youtube*. https://www.youtube.com/watch?v=FtOghXKKdUw

Somé, M. P. (1994). *Of water and the spirit: Ritual, magic, and initiation in the life of an African shaman*. Tarcher/Putnam.

Stiglitz, J. E. (2022). COVID has made global inequality much worse. *Scientific American*. Retrieved January 29, 2023, from https://www.scientificamerican.com/article/covid-has-made-global-inequality-much-worse/

Washburn, J. (2008). *University, Inc.: The corporate corruption of higher education*. Basic Books.

Williams, J. J. (2006). The post-welfare state university. *American Literary History, 18*(1), 190–216.

Zumper.com. (2020). *Santa Cruz, CA rent prices*. Retrieved January 29, 2023, from https://www.zumper.com/rent-research/santa-cruz-ca

Christopher Lang is a PhD candidate at the University of California, Santa Cruz who studies the relationship between race, consumer environments, waste, and community health. In his research, he considers the politics of disposability and uneven geographies of pollution made possible by state and corporate actors as well as convenience culture under racial capitalism. His prior schooling and work in marine conservation ultimately led him to explore commodity chains and consumer behaviors upstream that desecrate the ocean all while impacting human and nonhuman biota on land.

Emily Murai is a Lecturer at College Nine and the Environmental Studies department at the University of California, Santa Cruz. She teaches courses on critical reading, writing and thinking, writing-across-the curriculum/writing-in-the-disciplines (WAC/WID), and in the environmental humanities. She holds a Ph.D. in Human Geography from the University of Minnesota, an M.A. in Sociology from the New School for Social Research and a B.A. in American Studies and Psychology from UCSC.

Flora Lu is a Professor of Environmental Studies at the University of California, Santa Cruz and an ecological anthropologist specializing in conservation politics, environmental justice, political ecology and Indigenous resource governance. She served as the Provost of College Nine and John R. Lewis College from 2014 to 2023, during which time she helped establish the People of Color Sustainability Collective (PoCSC) and the Apprenticeship in Community Engaged Research (H)ACER. She was a Postdoctoral Fellow at Stanford University and an Assistant Professor in Ecology at the University of North Carolina at Chapel Hill before joining the UCSC faculty in 2008.

Appendix A: Self-Assessment for Diversity, Inclusion and Sustainability

Self-Assessment for Diversity, Inclusion & Sustainability:

Developed for UC Sustainability Officers from the workshop "Diversity, Inclusion & Sustainability: What can YOU do as a Sustainability Officer?" by UC Santa Cruz in Oct. 2016 (Elida Erickson, Dr. Rebecca Hernandez, Dr. Flora Lu, Adriana Renteria). Updated in 2022.

What can YOU do in your role as a sustainability professional in higher education?

1. LEARN MORE

Check off which one applies to you.

	Actively working on this	Thinking about this	Never thought about it
What is my current understanding of different cultural interpretations of how to care for the environment, and how can I learn more?			
What am I learning about my own cultural, ethnic and/or racial identities and how that intersects with my work and roles in society?			
How am I engaging in conversations with others about the complex issues of diversity, social justice, inclusion, power and privilege in society?			
What efforts am I taking to increase my own understanding around diversity-related terminology (ex: diversity, social justice, inclusion, equity, power, privilege, anti-racism, and more)?			
What can I do to learn more about how social justice, diversity, socioeconomic and environmental issues intersect in our global, local, and campus communities, and how that impacts our diverse student population?			

F. Lu, E. Murai (eds.), *Critical Campus Sustainabilities*, Sustainable Development Goals Series,
https://doi.org/10.1007/978-3-031-30929-8

Self-Assessment for Diversity, Inclusion & Sustainability:

What can YOU do in your role as a sustainability professional in higher education?

Developed for UC Sustainability Officers from the workshop "Diversity, Inclusion & Sustainability: What can YOU do as a Sustainability Officer?" by UC Santa Cruz in Oct. 2016 (Elida Erickson, Dr. Rebecca Hernandez, Dr. Flora Lu, Adriana Renteria). Updated in 2022.

2. Critically examine your organization
Check off which one applies to you.

	Actively working on this	Thinking about this	Never thought about it
Where does my organization tend to invest its financial resources, time and personnel?			
What communities are most benefited by my organization's work?			
How does my organization define sustainability? How does my organization define diversity?			
How valued would I say sustainability is within my organizational culture? What messages do I receive that reinforce that value?			
How valued would I say diversity is within my organizational culture? What messages do I receive that reinforce that value?			
In my position, do I have the ability to influence how diversity and/or sustainability is valued within my organization? What could that look like?			
Does my organization regularly encourage and support opportunities for professional development in diversity and/or sustainability?			
If my organization hires student positions, do I know whether these jobs are accessible to undocumented students? What can I do to help support undocumented students?			

Self-Assessment for Diversity, Inclusion & Sustainability:

What can YOU do in your role as a sustainability professional in higher education?

Developed for UC Sustainability Officers from the workshop "Diversity, Inclusion & Sustainability: What can YOU do as a Sustainability Officer?" by UC Santa Cruz in Oct. 2016 (Elida Erickson, Dr. Rebecca Hernandez, Dr. Flora Lu, Adriana Renteria). Updated in 2022.

3. Build relationships

Check off which one applies to you.

	Actively working on this	Thinking about this	Never thought about it
What type of campus and community events am I supporting and attending? How can I show up in new spaces?			
Through my role on campus, how could I engage staff, faculty and/or students doing diversity and equity-related work?			
Through my role on campus, how could I engage staff, faculty and/or students doing work at the intersections of sustainability, diversity and equity?			
Are there any existing relationships and partnerships that I can build on?			
Who can I connect with to engage in meaningful, transparent conversations and process what I am learning?			
How can I examine my personal approach to others in this work to ensure that I am authentic, accountable, and reflective when I have an 'oops' moment?			

Index

F. Lu, E. Murai (eds.), *Critical Campus Sustainabilities*, Sustainable Development Goals Series,
https://doi.org/10.1007/978-3-031-30929-8

The manufacturer's authorised representative in the EU is Springer Nature Customer Service Centre GmbH, Europaplatz 3, 69115 Heidelberg, Germany. If you have any concerns regarding our products, please contact ProductSafety@springernature.com

Printed and bound by CPI Group (UK) Ltd, Croydon, CR0 4YY
28/09/2026
02224075-0019